The
Dust
Bowl

R. DOUGLAS HURT

The Dust Bowl:
An Agricultural
and Social History

Nelson-Hall nh Chicago

LIBRARY OF CONGRESS CATALOGING IN PUBLICATION DATA

Hurt, R. Douglas.
  The Dust Bowl.

  Bibliography: p.
  Includes index.
  1. Agriculture—Great Plains—History—20th
century. 2. Great Plains—History. 3. Great
Plains—Social conditions. 4. Droughts—Great
Plains—History—20th century. 5. Dust storms—
Great Plains—History—20th century. 6. Depressions
—1929—United States. I. Title.
S441.H92   338.1′0978   81-4031
ISBN 0-88229-541-1 (cloth)   AACR2
ISBN 0-88229-789-9 (paper)

Copyright © 1981 by R. Douglas Hurt

Manufactured in the United States of America

10 9 8 7 6 5 4 3 2 1

For My Parents
Margaret and Ray Hurt

Contents

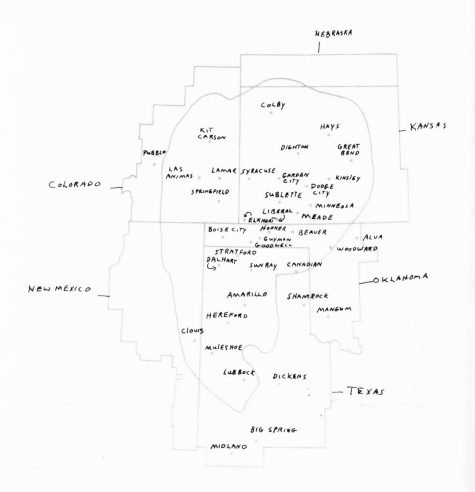

NEBRASKA

KANSAS

COLBY

KIT CARSON

HAYS

PUEBLO

DIGHTON

GREAT BEND

LAS ANIMAS

LAMAR

SYRACUSE

GARDEN CITY

KINSLEY

COLORADO

SPRINGFIELD

SUBLETTE

DODGE CITY

MINNEOLA

LIBERAL

MEADE

ELKHART

BOISE CITY

HOOKER

BEAVER

ALVA

GUYMON

GOODWELL

WOODWARD

STRATFORD

DALHART

SUNRAY

CANADIAN

OKLAHOMA

NEW MEXICO

AMARILLO

SHAMROCK

MANGUM

HEREFORD

CLOVIS

MULESHOE

LUBBOCK

DICKENS

TEXAS

BIG SPRING

MIDLAND

Preface

As a native of western Kansas, I grew up hearing my parents talk about the devastating black blizzards of the 1930s. The taped windows, the darkness at noon, and the all-permeating dust seemed nearly incomprehensible to me as a small boy. However, the dust storms of the 1950s gave me some indication of what the Dust Bowl had been like during the "dirty thirties." When a dust cloud rolled over my small home town during the "filthy fifties," my mother would invariably say, "I hope it's not going to start again." It was not until much later, after I had developed an interest in agricultural history, that I understood her apprehension.

This narrative of the agricultural and social history of the Dust Bowl is the result of my long interest in the Great Plains. I am particularly grateful, however, to Homer E. Socolofsky who suggested that I write the book. Professor Socolofsky's thorough knowledge of Great Plains history and his perceptive criticism of the manuscript have immeasurably benefited this study. I am also indebted to the Smithsonian Institution for granting me a Fellowship in the History of Science and Technology. With that aid, much of this research was conducted during my year-long residence at the National Museum of History and Technology. At the Smithsonian Institution, John T. Schlebecker guided the research. His insights were valuable beyond proper recognition. I am grateful as well to Mrs. Jeanne Claytor, who allowed me to use the papers of her father, John L. McCarty, which are still in her possession.

I am also indebted to many library and archival staffs for helping me to locate appropriate manuscripts, documents, and other materials. Douglas Helms at the National Archives generously led me through its pertinent collections. Jack Traylor and Pat Machaelis at the Kansas State Historical Society and Linda Bredehoft at the Amarillo Public Library, as well as their colleagues, were most helpful. I am thankful as well for the help that I was given at the Library of Congress, the National Agricultural Library, the Colorado State Historical Society, the Ohio Historical Society, Kansas State University, Texas Tech University, Ohio State University, and the Denver Public Library.

Portions of the manuscript were read by Charles Wood, Roger Lambert, and Katha Hurt. This study has benefited because of their help. Any faults or omissions are, of course, my own.

I am also thankful to Saundra Ball for typing the manuscript and to Chris Duckworth for preparing two of the maps. I am especially grateful to Mary Ellen Hurt for preparing the index. Their help greatly speeded the completion of this work.

CHAPTER ONE

*Prelude**

Sunday, April 14, 1935, dawned warm and dry on the
southern Great Plains. In Guymon, Oklahoma, worshipers
thronged to the Methodist Episcopal church to seek divine
deliverance from the drought then entering its fifth year. At
this rain service, the congregation heard Reverend R. L.
Wells proclaim that there would be little harvest unless rain
came soon. "Good rains within three weeks mean a harvest.
God rules all and our last resort is prayer." After church services in Amarillo, Texas, Sunday motorists flocked to the
roads to enjoy an afternoon outing in the spring air.[1]

Unknown to them, a high pressure system had moved out
of the Dakotas and into eastern Wyoming, silently lifting the
powder-dry soil of the Plains as it went. By early afternoon it
had created the most awesome "black blizzard" the people of
the Dust Bowl would experience in the 1930s. As the dust
cloud swept southward across eastern Colorado and western
Kansas, it extended in a continuous line from west to east.
From the height of a thousand feet the cloud boiled down-

---

* Portions of this chapter and of the following chapters were previously
published in R. Douglas Hurt, "Dust Bowl," from *The American West* magazine, July/August 1977. Copyright© 1977 by the American West Publishing Company, Cupertino, California. Excerpted by permission of the
publisher.

1

ward like the smoke of a gigantic oil fire, only to be met at the bottom by the rising columns of dust that it pushed before it. Inky black at its base, the cloud's color faded to a dark brown at the top as more sunlight permeated the dust. Some who saw it likened it to a great wall of muddy water. Hundreds of geese, ducks, and smaller birds flew in panic before it.[2]

The dust storm struck Dodge City, Kansas, with sixty-mile-an-hour winds at 2:40 P.M.; the nearly instantaneous and total darkness lasted for forty minutes. Semi-darkness followed for the next three hours, but by midnight most of the dust had passed. The storm traveled the 105 miles between Boise City, Oklahoma, and Amarillo, Texas, in one hour and fifteen minutes, stranding the Sunday motorists on the highways. During the height of the storm a Meade County, Kansas, woman dragged her rocking chair into the middle of the living room and sat down filled with a satisfying peace because the tape over the window frames was holding out almost every bit of dust. That was, she thought, "a condition under which almost any housewife could have died happily." Occasionally, she turned off the lights to see whether the windows were admitting any light, but all she could see was a dark mass of soil pressed tightly against the panes. The dust hung like a curtain with no visible motion. Most people who experienced the storm found it terrifying. For one, "It seemed as if it were the end of all life;" another likened the darkness to "the end of the world;" and still another professed, "The nightmare is becoming life."[3]

Although the boundaries of the Dust Bowl were never precise, its general location encompassed that part of the southern Plains where the drought and wind erosion hazard were the worst—a 97 million acre section of southeastern Colorado and northeastern New Mexico, western Kansas, and the panhandles of Texas and Oklahoma. This area extended roughly four hundred miles from north to south and three hundred miles from east to west with the approximate center at Liberal, near the southwestern corner of Kansas. Wind erosion was most severe within a hundred-mile radius of that town. The outline of the Dust Bowl changed from year to year, depending upon the amount of annual precipitation; however, the "blow area" never covered the entire 97 million acres which were potentially subject to severe wind erosion. Instead, the Dust Bowl reached its greatest extent from 1935 to 1936 when it covered about 50 million acres and was concentrated largely in southwestern Kansas.[4]

The frequency and severity of the dust storms during the 1930s became major news items, and journalists from various newspapers and magazines were sent to the Plains to describe the awesome scene. Following the April 14, 1935, black blizzard, Robert E. Geiger, an Associated Press reporter, released a series of articles, the first from Guymon, Oklahoma for the *Washington (D.C.) Evening Star.* In his story, Geiger inadvertently but appropriately used the term "dust bowl." Geiger ignored this term in his next two articles and in his last story from the area, he referred to the region as the "dust belt." The public and the Soil Conservation Service, however, adopted the term Dust Bowl almost immediately and used it when referring to this windblown, drought-stricken area.[5]

By the spring of 1935 the people of the southern Great Plains were entering their fourth year of severe dust storms, and even though the "blow months" of February, March, and April were the worst they had ever experienced, drought and dust storms were not new to the Plains. After the Pleistocene era, the Great Plains had experienced two prolonged drought periods (the Boreal from 8000 B.C. to approximately 5000 B.C. and the Sub-Boreal from 2000 B.C. to approximately 600 B.C.) during which time less precipitation fell there than anywhere else farther east. In addition, the study of tree rings from red cedar and yellow pine trees in western Nebraska indicates that during the 748 year period prior to 1958, twenty-one droughts occurred which lasted for a duration of 5 or more years, with recurrence every 35.7 years. During that time, the average drought lasted 12.8 years, and a drought of 10 or more years came every 55.6 years.

Droughts, then, have occurred for centuries in the Great Plains and will continue to recur as long as man is unable to control the climate.[6]

With these droughts came dust storms. They were a natural phenomena of the Great Plains. Indeed, ever since anyone can remember, dust has always blown there. In October 1830, Isaac McCoy, a Baptist missionary commissioned to survey the land which had been assigned to the Delaware tribe, ventured into eastern Kansas and encountered a dust storm that made it impossible for him to follow the trail of the surveying party which had preceded him by only a few hours. The following month, McCoy experienced another dust storm which, mixed with the ashes from a recent prairie fire, reduced visibility to the extent that he could not distinguish objects farther away than three or four times the length of his horse. "The dust, sand, and ashes," he wrote, "were so dense that one appeared in danger of suffocation."[7]

In May 1854, a new settler on the eastern fringe of the Great Plains reported that friends warned of high winds and dust in the early spring, and, in late January 1855, the editor of the *Lawrence, Kansas, Free State* noted that strong winds and black dust from the burned prairie penetrated every home. By April of that year blowing dust was so bad that another editor reflected:

> The high winds which have prevailed in this vicinity for the last few weeks, accompanied with heavy clouds of dust, have no doubt been a source of very great annoyance to strangers who have been on a visit to the Territory, as well as to the citizens. Whether those winds are common to Kansas in the spring we are not informed, probably they are; but the dust, which is the most annoying, is a resultant of the burning of the prairies, and will not exist after the annual fires have abated. Neither will they harm us after the grass shall get high enough to prevent the wind from taking up the surface, and hurling it with so much force through the atmosphere.
> We are frank to confess that we have felt more inconvenience from the wind and dust, since our arrival in Kansas,

than from any other source. Our houses are all open, and the wind whistles in at every crevice, bringing along with it a heavy load of fine particles of charcoal, ashes, etc., and depositing it on our type, paper, library, furniture, and in fact not regarding our dinner, but liberally covering it with condiment for which we have no relish.[8]

Drought became severe in 1860 and brought dust storms that sometimes reduced visibility to several yards. This drought ranks in severity with that of the 1930s. From 1860 to 1864, the driest years, Leavenworth, Kansas, on the eastern edge of the Great Plains, averaged 24.3 inches of precipitation annually—10.8 inches below normal. The driest year was 1864, when this city received only 9.2 inches of precipitation. During the 1930s, Leavenworth's average precipitation was 31.7 inches. The drought stunted vegetation and the prevailing winds lifted the soil into the air. The dust storm on April 5, 1860, was severe, and the editor of the *Fort Scott* (Kansas) *Democrat* remarked that it was "by far the most disagreeable we ever experienced. For the space of half an hour the cloud of dust was so intense that it was impossible to distinguish objects at the distance of a dozen yards."[9]

During the following decade the drought was serious in western Kansas. The *Junction City* (Kansas) *Union* reported on February 19, 1870, that "a great deal of Kansas is not located where it used to be. Some of it we have no doubts is located in South America, while some covers the British possessions." In April 1872 the *Wichita Eagle* reported, "Real estate for sale at this office, by the acre or bushel. We have no disposition to infringe upon the business of our friends down the street, but owing to the high winds and the open condition of our office, and not being ready for interment just yet, necessity compels to us[sic] dispose of the fine bottom land now spread over our type and presses."

At Holton two years later, after prairie fire further reduced the drought stunted grasses, "The wind blew almost a hurricane and such immense clouds of dust filled the air that very few people ventured out. . . . The wind was fearful, the dust intolerable." More dust storms occurred in 1878, one of which prompted the *Abilene* (Kansas) *Gazette* to note that "a furious wind storm passed over the city last Friday [August 2] evening, filling the air so thick with dust that it was difficult to distinguish objects ten yards away."[10]

Probably the worst dust storm of the nineteenth century swept across the Great Plains on March 26–27, 1880. The storm extended from Las Cruces, New Mexico, to Iowa and eastern Missouri. The dust almost obscured the sun at 10:00 A.M. in Leavenworth, Kansas, and drifted soil 1 to 2-1/2 feet deep at Howard, Nebraska. On March 27 the atmosphere from eastern Kansas across the entire state of Missouri, except for the extreme southeastern part, was filled with a fine grayish dust which darkened the sunlight and decreased visibility to a hundred yards. The dust storm gave everything a "sickly, indescribable melancholy appearance." During the storm, the wind blew hard from the west and northwest and caused much property damage.

The *Wichita Eagle* reported that this dust storm, though not unlike the dust storms of each spring, struck with a greater force than usual and covered the south-central Plains with a "yellowish impenetrable fog." Static electricity came with the storm and hindered the operation of the telegraph. The editor noted, "These storms generally occur about the equinox and are not dangerous but exceedingly disagreeable." The more educated citizens of Leavenworth likened the dust storm to Bulwer's description of the day preceeding the fall of Pompeii in 79 A.D. The *Kansas City* (Missouri) *Journal* reflected that such sand storms, while rare in eastern Kansas and Missouri, were common in the plains of western Kansas and eastern Colorado. But, they were not dangerous, just inconvenient. The *Winfield* (Kansas) *Daily Courier* referring to this dust storm reported, "No wonder the soil of Butler County is growing richer by the year. If Cowley County's soil continues to move northward as rapidly as it did Tuesday, Butler will soon be one of the best counties in the state."[11]

One Kansan offered an explanation for the severity of this dust storm. He attributed the dust to small particles of dry grass which the wind lifted and carried eastward. His theory was that in "the absence of snow during the winter season, the dry grass blades, especially the edges, break into these dusty fragments, and find lodgings beneath the tops of the grass, and remains there unless disturbed by the wind."Another theorist suggested the dust storm was the remnants of a shattered or atomized meteor.[12]

A little over two weeks later, a housewife in Ogden, Kansas, swept eleven pounds of dust from her kitchen and hall following another dust storm and estimated that at least a hundred and ninety pounds of dirt had blown into her home. The *Wichita Eagle* complained on April 15:

> It may be asserted here and now that Kansas as a paradise has her failings, not the least of which is her everlasting spring winds. If there is a man, woman or child in Sedgwick County whose eyes are not filled with dust and their minds with disgust, he, she, or it must be an idiot or awful pious. From everlasting to everlasting this wind for a week has just sat down on its hind legs and howled and screeched and snorted until you couldn't tell your grandfather from a jackass rabbit. And its sand backs up its blow with oceans of grit to spare. We saw a preacher standing on the corner the other day with his back up, his coattails over his head, and his chapeau sailing heavenward, spitting mud out of his mouth and looking unutterable things. He dug the sand out of his eyes and the gravel out of his hair, and said nothing. It wouldn't have been right. But we know what he thought. As for our poor women, weighted down with bar lead and tracechains as their skirts are, their only protection from rude gaze is the dust which fills up the eyes of the men so that they can't see a rod further than a blind mule. Dust, grit, and sand everywhere—in your victuals, up your nose, down your back, between your toes. The chickens have quit eating gravel—they absorb enough sand every night to run their gizzards all next day. Out of doors people communicate by signs. When they would talk they must retire to some room without windows or a crack, pull out

their ear plugs and wash their mouths. The sun looks through
fathoms of real estate in a sickly way, but the only clouds
descried are of sand, old rags, paper and brick bats. We
haven't done the subject justice, but we didn't expect to when
we started out, but it blows, you bet.[13]

Another big dust storm swept across most of west-central
Kansas on April 18, 1880, and piled dirt an inch deep in some
houses. In some places the wind drifted the soil like snow.
This dust storm struck on a Sunday morning about the same
time that a minister in the Wichita area was telling his congregation about the reward of the impenitent. The *Wichita Eagle*
reported, tongue in cheek, "He had got into the fire of his
peroration and the smoke was ascending forever and forever
when forever out went the gable end of the meeting
house. . . . Stopping short, he looked up through the hole in
the roof and then to the faces of his frenzied audience, and
earnestly remarked, 'My dear sinners, why waste further
words to describe that awful place when you can get a better
idea of it by simply sticking your heads out of doors.'" Kansans then, as well as Dust Bowl residents later, kept their
humor about the dust storms.[14]

Dust storms recurred periodically during the 1880s. In late
September 1881 near Solomon, Kansas, blowing dust was
making life "eminently disagreeable" and "the man who
refrained from profanity and growling was a rare exception
and hard to find." In January 1883, the areas around Las
Animas, Colorado, and Fort Union, New Mexico, had four
dust storms. In March, Kansas reported three storms—one of
which was "stupendously suffocating all day." The following
month Nebraska reported five sand storms. On April 19,
1883, the *Kinsley* (Kansas) *Graphic* commented on a recent
dust storm:

Kinsley was visited on Friday last by one of the most ungentle
zephyrs that it has been our misfortune to experience since a
resident here. A perfect cloud of dust and sand filled the air

and dusted in every crack and crevice of the buildings, and the
unlucky pedestrian who was compelled to be abroad absorbed
the full peck of dirt that is allotted to each one's life, and what
his or her stomach would not hold was stowed away in their
ears, eyes and clothing. The flies laid low; the dogs crawled
into the cellars, and the birds nestled closely wherever shelter
could be found. Irrigators as well as anti-irrigators prayed loud
and continuously for water—it mattered not how or by what
means it was obtained—and their prayers were answered as
the day wound up by a heavy rain and hail storm. We suppose
it is necessary that we should occasionally have these little visi-
tations or we would not know how to appreciate the delightful
climate with which we are blessed, as there is certainly no
other good object attained.[15]

On April 16, 1886, the *Abilene* (Kansas) *Gazette* remarked
that the recent Kansas zephyrs had blown with unusual force
from the south. Yet, one woman asserted that the wind and
accompanying dust nicely offset the "gloomy, rainy days, and
the mud so common and so very disagreeable in the Eastern
states." She preferred dust on her furniture and carpets to
mud. Some boosters attested that "were it not for the agita-
tion of the atmosphere by the winds Kansas would not be, as
it is, the healthiest state in the Union." In October, sixty-
mile-per-hour winds raised heavy clouds of sand and dust and
completely obscured the sky at North Platte, Nebraska.
From January through April 1887, sand storms periodically
struck Midland and Abilene, Texas. Farther to the north at
Wolsey, Dakota Territory, a severe sand storm struck on
April 2, 1888. There, "the wind was very high during the day
and drifted the sand three inches in places." Near Yankton,
Dakota, this storm obscured the sun and made artificial light-
ing necessary in stores and homes. On February 7, 1889, the
Salina (Kansas) *Journal* reported that "howling winds" several
days earlier had "lifted the surface of the earth into the air."[16]

In 1892, a serious drought occurred in southwestern Kansas, and by 1894 it was general across the state. The average
precipitation for the western third of the state for 1893 and

1894 was 11.9 and 12.2 inches respectively—about 63 percent of normal. Once again the dust storms came with the
drought. On August 13, 1892, at Dodge City, a dust storm
limited visibility to not more than a hundred and fifty feet.
The dust storm at Meade on April 13, 1893, drifted the soil
like snow and caused one newspaperman to note, "During
the last ten days real estate has probably been as high—in the
air—in Kansas, as any time in the history of the state."
Another blinding dust storm in April deposited so much soil
on office furniture in Dodge City that everything looked as
though it were being prepared for a hot bed. In mid-May the
following year, high winds made the dust "terrible" in that
city.[17]

One Nebraska farmer recalled a serious dust storm which
struck in the spring of 1893. Although dust storms had been
occurring almost daily all spring, a violent dust storm swept
over his farm near Stratton about nine thirty one morning. By
noon the storm had increased in intensity and reduced visibility to ten or fifteen feet. Because these storms were frequent,
local farmers made a habit of staking down everything that
might blow away. After the storm, however, he found a
nearby timber claim littered with fence boards, barrels, boxes
and enough shingles to roof "half the county." The storm had
blown in from the southwest and had driven sand through the
air with such velocity that it pitted the bay window on the
south side of the house to the extent that vision through it
became difficult. Venturing outside during the storm was
unhealthy if not dangerous, because as one observer put it,
"Every time we went out of doors in this storm, it was necessary to wear rags of some kind over our faces to keep the sand
from literally cutting the skin off our bodies." Although this
dust storm was bad, he avowed worse ones had occurred in
western Kansas.[18]

In mid-September 1893, a severe sand storm struck the
Cherokee Outlet in what is now northern Oklahoma. A rapid
succession of sand storms followed during the next two
weeks, with visibility often less than twenty feet. Similar

storms plagued the area for the remainder of the decade. A dust storm struck Stillwater, Oklahoma, on January 20, 1895. The wind, with gusts up to fifty-five miles per hour, lifted columns of dust a thousand feet high and sifted it into every household crevice. Another dust storm swept over the Oklahoma Territory near South Enid on March 19, 1895. The wind blew at eighty miles per hour from 4:00 P.M. until 2:00 A.M. suspending travel and damaging property. When the storm abated, several inches of dust covered wheat and vegetable crops in the sandy lowlands. At El Reno, Oklahoma, an early April sand storm nearly stopped business in addition to impeding travel. During the storm, "buildings could not be seen fifty yards and the sand was scattered along as though sown broadcast from a great hand."

The April 12 storm at Larned, Kansas, drifted sand six feet deep along the railroad tracks as far west as the state line. This storm smothered horses and cattle "by the score." Approximately five thousand horses and cows were killed between Larned, Kansas, and Lamar, Colorado. One Gray County, Kansas, farmer lost thirty-two head out of a herd of forty-six cattle. On April 15, a sand storm struck southern Kansas, Oklahoma, and the Texas Panhandle. The results were so terrible that "the superstitious thought the astronomical conditions which, it is said, are now repeating themselves for the first time since the death of Christ, had something to do with it."[19]

Dust storms also plagued eastern Colorado in April 1895. Several of these storms were so intense that trains were stalled until crews with snow plows and scoop shovels could clear the cuts from the drifted sand. On April 9, a fierce sand storm in northeastern Colorado and southeastern Wyoming lacerated the workmen clearing the railroad tracks and killed an estimated 20 percent of the range stock. Many cattle drifted over a hundred miles before perishing in the storm. Several days later, a dust storm which gave a pink color to a Colorado snowfall brought "Egyptian darkness" to western Oklahoma and drifted sand six feet deep in parts of western

Kansas. On May 10, one of the worst dust and sand storms in years swept across Oklahoma and into Kansas. At Pittsburg, Kansas, the storm raged all afternoon, "caking everything in its path." The dust darkened the sky so much at Guthrie, Oklahoma, that some people feared a tornado and fled to the safety of their cyclone cellars. The following month, one Kansas journalist happily wrote: "One of the seven wonders of the world. We haven't had a dust storm for a week."[20]

In 1895, the dry autumn and winter killed much of the wheat crop; in the following spring many farmers began to plow the crop under. As a result, the wind blew the soil, and on March 28, 1896, the first major dust storm of the season struck Dickinson County, Kansas. The wind blew hard out of the southwest. As the dust moved through the air, it made the residents "prone to indulge in unparliamentary language." In March 1899, the wind blew so hard near Prudence, Oklahoma that farmers quit trying to work their fields for fear the oats sown would blow out of the ground before it could sprout. Here, dust drifted eighteen inches deep along the fields; in many places the wind removed the top soil to the hard pan, that is to the depth of the plowing. In late April, a dust storm struck eastern Nebraska; at 2:30 P.M. a light was needed to read by. One Nebraskan wrote: "The weird sky, the rapidly rolling mass of yellowish clouds to the northeast, and the falling dust had brought us all out to watch, ready for a hail storm or a tornado and a dive for the cave." A little rain accompanied the storm and turned the dust to mud. Muddy rain also fell over eastern Dakota and western Iowa.[21]

In Kansas a precipitation deficiency recurred in November 1900. By the following year, the drought was serious in all parts of the state, and high temperatures during the summer months intensified the problem. The next year, eastern Kansas received a little over 25 inches of precipitation—nearly 10 inches below normal. The central third of the state received 21.3 inches or some 5 inches below normal. In western Kansas, only 17 inches of precipitation fell for about a 2 inch deficiency. Again, the drought brought dust storms which increased in number over the central plains from 1901 to 1904.[22]

In 1910, a deficiency of rainfall occurred in the south-central Great Plains and lasted until 1918. From 1910 until the end of 1914, Kansas was almost as dry as it was during the five driest years of the 1930s. In June 1911, the state averaged only .64 inches of precipitation, making it the driest June on record. Kansas averaged only .47 inches of precipitation in August 1914—also the driest August recorded. That month was also the hottest on record until August 1936.[23]

As a result of the drought, the dust began to blow in Thomas County, Kansas, during the spring of 1912. Farmers near Colby reported that a strip of land northeast of town about fifteen miles long and five miles wide had "blown out." The wind blew out the winter wheat which was already several inches high. Not a sprig of vegetation remained to hold the soil, and the ground was as hard as a city street; the top soil blew away to the hard pan. The dust storm which this blowout created lasted only several hours, but it was severe enough to make travel hazardous and it forced residents to light lamps.[24]

Dust storms of lesser intensity continued day and night for the next three or four days. Again the soil drifted and home owners shoveled pathways through it. In some places the soil formed great drifts from five to twenty feet deep and covered roads and fences. The blowout area had the appearance of a "bleak brown desert of unbroken earth." A journalist from the *Kansas City Star* drove through the area during a June dust storm and reported that only the dim outline of the telephone poles enabled him to stay on the road.[25]

The drought caused the wheat crop to fail in 1911 and 1912, and little root system remained to hold the soil against the wind. In addition, farmers pulverized the fields with their plows as they turned under the failed crop. The cumulative result was the creation of a fine, powder-dry soil exposed to the wind, and more dust storms followed. Some people argued that dust storms were actually beneficial to agriculture. Since the wind had blown about six inches of top soil away, the dirt on the surface was virgin soil and therefore more productive. Farmers as well as others would make this same false assumption during the 1930s.[26]

Because of successive crop failures, most farmers lacked the money to undertake a conservation program that was necessary to bring their lands under control. In the spring of 1913, Thomas County farmers appealed for aid to the Rock Island and Union Pacific railroads, which ran through northwestern Kansas. The railroads agreed because it was in their best interests to help to bring the soil under control and to stabilize agriculture in the area. This would enable the railroads to continue profiting from hauling grain and cattle out and from transporting in needed seed, fertilizer and equipment for the farmers. The Rock Island contributed $4,000, the Union Pacific gave $1,000, and wholesalers and jobbers in Topeka and Salina contributed $1,500 for conservation expenses. With this money, H. M. Cottrell, the agricultural commissioner for the Rock Island, hired the farmers to "list" the blown out area. This policy not only gave the farmers some much needed financial relief, it also helped check the soil blowing. (The lister is essentially a double moldboard plow which splits the furrow and turns the slice each way.) Cane and kafir corn were planted in the listed rows. The deep listed furrows helped to catch the blowing soil, and by the time the furrows drifted full, the cane and kafir corn were tall enough to help check further soil movement. The land was not blank listed—that is, listed solidly across a field—because of the expense. Rather, farmers listed five parallel rows and then left a space of fifty or sixty feet and began listing again. This method proved satisfactory for substantially checking soil blowing. Thomas County commissioners, the sheriff, and local volunteer "dust committees" helped the farmers with the work. One local farmer using this method lister-planted seventy acres of drought-resistant milo, and harvested fifteen bushels to the acre. With the selling price at $1.50 per bushel, he received $22.50 per acre from his crop on what would otherwise have been a barren, windblown field.[27]

By the following spring, the blowout area had been largely reclaimed. Local rains were sufficient to make good wheat, barley, and corn crops. Some wheat looked good enough to make thirty-five bushels per acre. With the dust settled and the soil held against the wind, farmers regained their confidence and began to purchase new machinery for the coming harvest.[28]

Between 1909 and 1914, there were fifty-six reports of drifting soil in Oklahoma. Near Hooker, in 1909, some farmers lost two-thirds of their wheat crop because of wind erosion while others in western Oklahoma abandoned their lands. By 1910 most of western Oklahoma was farm land, except for the extreme west end of the panhandle which remained principally in grass. Here as elsewhere in the south-central Plains, farmers did not practice soil conservation. Planting and cultivation of wheat, cotton, and corn invariably left the soil exposed to wind and water erosion part of the time. As the sandy soils became pulverized from plowing, the land became more susceptible to blowing. Consequently, less than forty years after the opening of the Oklahoma Territory for settlement, a serious soil erosion problem developed. Indeed, the newly settled farm land of Oklahoma was one of the most seriously eroded sections in the nation.[29]

In retrospect, dust storms in the southern Great Plains, and indeed, in the Plains as a whole, were not unique to the 1930s. Drought, lack of vegetation, and wind have caused the dust to move since the formation of the Plains. The elimination of any one causal element, though, will significantly reduce or eliminate dust storms. When all three elements are present, however, the dust blows. During the early nineteenth century and before, when buffalo were the primary occupants of the Plains, drought and prairie fires destroyed the native grass and exposed the soil to wind erosion. Later in the nineteenth and early twentieth centuries, however, other factors contributed to dust storms—notably man's inhabitation of the southern Plains and the adoption of a new agricultural technology.

CHAPTER TWO

Causes of the Dust Bowl

By driving south from Dodge City, Kansas, on Route 283 to Minneola, and then by turning southwest on Route 50, a traveler passes through Meade and Liberal before reaching the Oklahoma Panhandle where Hooker, Guymon, and Goodwell appear in a row before the long, desolate stretch of road to Stratford, Texas. Then, by turning north on Route 287, a driver can reenter the Oklahoma Panhandle on his way to Springfield and Lamar, Colorado. Such a trip through the heart of the Dust Bowl takes less than a day. No matter which season of the year, the countryside is quiet, and one's attention focuses at first on the monotonous, level, treeless plain. Looking closer, one sees smooth, undulating farm country broken only by occasional hills, creeks, and ravines, which gradually rises from 2,000 feet to over 5,000 feet in elevation. Within this relatively isolated region, the composition of the soil, the severity of the climate, and the settlement of man were responsible for the creation of the Dust Bowl.[1]

Dust Bowl soils are generally rich in potassium, phosphorous, and nitrogen, but their composition varies from a deep, heavy loam to a shallow, structureless sand. Three major soil groups dominate the southern Plains—chernozem in the eastern half, southern dark-brown soils in the western half, and the brown soil of eastern Colorado and northeastern

New Mexico. The chernozem soil is the most productive and responds well to wind erosion conservation measures. The southern dark brown soils are more sandy than the chernozem and are more subject to severe wind erosion if not properly controlled. Southern dark brown soils are not well suited for wheat production, but prior to the 1930s, Dust Bowl farmers used them extensively for that purpose. The brown soils are relatively sandy, and they are best suited for soil-holding row crops or grasses, but they too were heavily planted in wheat before the soil began to blow. At least 50 percent of the crop land in this soil type needed to be regrassed during the 1930s to help to stabilize the land and to prevent further blowing. Overall, conservationists recommended the return of 6 million acres to grass out of 32 million cultivated acres in the Dust Bowl. Only 26 million acres were considered arable, provided the proper soil conservation measures were applied.[2]

These Dust Bowl soils were primarily created from friable calcareous and sandy clays washed down from the Rocky Mountains by melting glacial waters. Some of the soil was wind deposited and it continued to blow until being covered with vegetation. As a whole, Dust Bowl soils have a characteristic known as "flocculation" which causes minute soil particles to cling together in clusters similar to fish eggs. Small channels separate the soil particles and facilitate the movement of moisture to the subsoil. Long periods of drought destroy flocculation, particularly when the soil is intensively cultivated. When flocculation is destroyed, the soil breaks down into tiny dust particles which are easily windblown. Flocculation is not restored until normal precipitation returns. Even then, two or three years of favorable precipitation are required to restore the soil to its normal structure. In the meantime, the soil blows and further breaks up the channels between other particles. As a result, moisture cannot readily penetrate it, rainfall runs off, vegetation fails to grow, and more dust blows, further covering vegetation and damaging the land.[3]

The climate of the southern Plains also contributed to the creation of the Dust Bowl. Winters and summers are severe; the variation between the maximum summer and the minimum winter temperatures often exceeds 100°F. The average growing season varies from one hundred thirty-five days in northern Colorado to two hundred and fifteen days in the southern part of the Texas Panhandle. Strong winds are common in the early spring and throughout much of the summer. The humidity is low, and the evaporation rate is high. Precipitation averages less than twenty inches annually, and it is irregularly distributed. Buffalo and gramma grasses grow well in this semiarid environment, but drought can damage even these hardy perennials if they are not managed properly. Prior to the 1930s, they were not; few farmers understood the need for grassland management.[4]

A harsh climate, different soil types, and tough prairie sod slowed the settlement of the Great Plains. In 1862, Congress passed the Homestead Act to stimulate the settlement of the Plains. The Homestead Act allowed each head of a household to claim 160 acres and receive free title to it provided he lived on the land for five years, cultivated a portion of it, and made improvements. A decade later in 1873, the Timber Culture Act permitted settlers to claim a quarter section (160 acres) of land, if they agreed to plant 40 acres of trees (later amended to 10 acres). The federal government assumed that tree planting would increase rainfall and make farming in the Plains less hazardous. Four years later the Desert Land Act allowed ranchers to claim 640 acres provided they agreed to irrigate it within three years. These federal land acts, among others, helped to lure settlers into the Plains, but the railroads aided too. The federal government granted railroad companies large tracts of land to defray the costs of building transcontinental lines. These lands sold for an average of $2.50 to $10.00 an acre. Although these lands were more expensive than the $1.25 an acre commutation cost under the Homestead Act, many settlers bought railroad lands and began farming.[5]

Still, from the end of the Civil War until 1886, cattlemen dominated the Great Plains. Settlement gradually expanded, but many people believed the Great Plains were uninhabitable or at least unsuitable for extensive agriculture. As late as 1900, the U. S. Geological Survey noted that much of the southern Great Plains was "nonagricultural," because of insufficient rainfall and lack of irrigation potential. However, the geological surveyors considered the region to be an ideal pastoral region because of the abundant shortgrass cover. Even so, artificial stock-watering points were needed, provided that wind mills could adequately pump water from the strata below.[6]

Superficially, the report of the U. S. Geological Survey seemed correct. Because the early settlers in the southern Great Plains owned few implements, inadequately understood which crops were most suitable, or did not know how to preserve soil moisture, many farms failed during the periodic droughts. Since government aid was not provided before the New Deal to enable destitute farmers to remain on their lands, the area was alternately occupied during times of high precipitation and abandoned when rainfall declined. Over time, southern Plains farmers became more knowledgeable about crop varieties and were less likely to emigrate during dry periods.

During the 1880s, the agricultural settlement of the Great Plains began on a large scale. With the Plains tribes removed to the reservations and with precipitation above average, homesteaders rapidly occupied the region. Between 1886 and 1910 an increasing number of farms were established in the southern Plains. Drought from the late 1880s to the mid-1890s caused a temporary setback, but after the turn of the century farmers intensively plowed the range. By 1910, farmers claimed almost all of the southern Plains, and wheat farming was the main activity. As farms spread across the region, more sod was broken and more soil was exposed to oxidation and wind erosion.[7]

At first, wind erosion was negligible on the newly broken lands, though some soil blowing did occur. In the Oklahoma Panhandle, wind erosion became a serious problem within less than a decade after plowing the sod. Economic necessity made the problem worse. Settlers needed money, and cash crops such as wheat provided it. When crops failed during a drought, the wind blew the barren soil. Many farmers could not afford to practice soil conservation, and some did not think it was necessary; others either did not know how or had been similarly negligent on their farms back east. However, some farmers did understand the need for soil conservation. They used plows to roughen the soil to slow the movement of the wind, and they planted drought-resistant grain sorghums. These farmers, however, were in the minority.[8]

Careless or not, settlers steadily increased the number and size of the farms in the southern Plains. In 1890, only 5,762 farms and ranches existed in a twenty-two-county area in the Dust Bowl portions of Kansas, Colorado, and Texas. These farms averaged 256 acres with only 90 acres in crops; 96 percent of the range was still unbroken. Twenty years later, however, the agriculture of the southern Plains began to change as farmers shifted to crop farming. By 1910, 11,422 farms were located in the twenty-two-county area, and farmers were plowing more sod than ever before. Cattle raising was still important, but emphasis on wheat meant financial ruin during severe droughts because there were few other crops to rely upon for income. When the wheat crop failed, more land was subject to wind erosion. For the most part, though, the southern Plains received adequate precipitation, the wheat grew, and farmers expanded their acreage with newly purchased implements.[9]

Farming in the Great Plains required implements that would speed tillage and harvesting and thus increase the number of acres one man could work. Indeed, a new technology had been necessary to make farming possible in the Great Plains. The tough sod required a sharp steel plow that scoured well, cut the root system, and turned over the sod. Furthermore, the nearly level terrain permitted the use of large machinery such as headers, steam engines, and threshing machines which in turn allowed farmers to plant more wheat to help compensate for drought losses.[10]

During the Civil War, the sulky or riding plow was perfected, and Kansas farmers used it widely during the 1870s. When the sulky plow was fitted with a second share, one man could plow substantially more than he could with a single-share sulky or even a walking plow. The lister plow was also an important implement which Plains farmers used to expand their operations. Most listers had a seed canister and drilling device attached. Although listing required increased draft power, it did several jobs at once since it plowed, planted, and covered the seed, all in one operation. Listing allowed farmers to make the best use of soil moisture, because the deep furrows helped to retain precipitation. The lister left the ground in a rough, cloddy condition and helped retard soil movement. These plows were relatively cheap (one cost about $35 in 1888), and they were well suited for Great Plains agriculture.[11]

After plowing, Plains farmers used endgate seeders, grain drills, rotary-drop corn planters, and check-row corn planters. These implements enabled more uniform planting and contributed to higher yields than could be obtained by hand broadcasting. Wheat farmers preferred the hand-rake and self-rake reapers and the Marsh harvester for cutting grain. Reapers cut the wheat and deposited it in piles on the ground. Binders walked behind the machine and tied the straw into bundles. The Marsh harvester enabled a man to ride on the reaper's platform and bind the grain there, thus saving time and speeding the harvesting process. Self-binders appeared in the late 1870s and were in general use with horse-powered threshing machines. Headers also appeared in the late 1870s; these large, cumbersome machines cut only the heads and left the straw. A header did a better job than a binder and cut a wider swath, but the grain still had to be threshed and winnowed. Nevertheless, the harvesting operation was speeded up.[12]

Steam power came into increasing use in the mid-1870s and by 1885 the traction steam engine was a success. Steam power enabled improved threshing efficiency, and considerable Dust Bowl land was broken with steam tractors and gang plows. Complete mechanization awaited the perfection of the gasoline tractor and with it the combine. In the meantime, Great Plains farmers plowed, seeded, and harvested as much land as they could afford in time and money. Continued increases in farming efficiency decreased costs and encouraged the purchase of more implements and land. In twenty-seven southern Plains counties, the value of farm implements and machinery rose from $4.6 million in 1910 to nearly $16.7 million in 1920. During that same time, the average farm increased in size from 465.5 acres to 771.4 acres. As farm units became larger, the amount of land in farms also rose from 10.0 million acres to 17.6 million acres or from 38.9 percent to 68.4 percent of the total land area. In addition, the number of cattle raised also increased from 506,583 head to 894,859 head.

In 1910, farmers harvested 970,344 acres of corn, oats, wheat, barley, rye, and sorghum in the twenty-seven-county survey area. In 1920, the harvested acreage for these crops totalled 2,285,296 acres. During that time, wheat lands expanded from 465,653 acres to 1,188,934 acres. All of these increases meant less grass remained to protect the soil. Increased tillage and expansion of farm holdings were also made at the expense of wise land-use practices.[13]

With the outbreak of war in Europe, wheat prices soared in the Dust Bowl states from an average of $0.91 per bushel to $2.06 per bushel by 1917; prices remained above $2.00 until 1920. When the United States entered the conflict in 1917, southern Plains farmers responded to the "Wheat will win the war!" campaign by planting still more. Rainfall remained sufficient to allow, if not encourage, that expansion. After the war, wheat expansion continued, and grazing land appraised at $10.00 an acre increased ten times in value when it was planted in wheat. The proceeds of a good crop year might equal the profits received from a decade of stock raising. When wartime prices collapsed in the early 1920s, plainsmen broke more sod (largely with the newly adopted one-way disk plows) in order to plant still more wheat to offset the economic loss. New technology, war, and depressed prices stimulated Great Plains farmers to break 32 million acres of sod between 1909 and 1929 for new wheat lands. In the southern Plains, wheat acreage expanded 200 percent between 1925 and 1931; in some counties this expansion varied from 400 to 1000 percent.[14]

The increased use of power machinery also stimulated this expansion, and southwestern Kansas offers an excellent example of that influence. In 1915, when the state board of agriculture first reported tractor statistics, only 286 tractors were located in that twenty-five-county area. Five years later, the number had increased to 1,333. By 1925 the number of tractors totalled 3,501, rose to 9,727 in 1930, and peaked at 11,655 in 1934. Combines were first used in that same area about 1917, but only 719 were in use by 1923. Two years later, though, the number of combines totalled 1,085, reached 6,083 in 1930, and rose to 7,724 in 1932. These combines reduced harvesting costs from more than $4.00 per acre by using a binder and threshing machine to about $1.50 per acre. Combines also increased the yield per acre because less grain was wasted.[15]

As farmers bought more tractors and combines, the percentage of the total land area seeded to wheat also increased. A farmer could now produce a bushel of wheat for each three minutes of labor, and wheat farming resembled factory production in both speed and quantity. In 1915 southwestern Kansas had 9.9 percent of the total land area in wheat. Five years later, it totalled 13.6 percent. It rose to 17.8 percent in 1925, and reached 38 percent in 1931. In Ford County, farmers seeded about 300,000 acres or half the county to wheat annually between 1917 and 1923; nearly 100,000 more acres were planted from 1924 to 1931. Some of that expansion resulted from a decrease in the acreage of other crops, but much of it resulted from the breaking of new sod. In 1918, for example, Ford County had 278,000 acres of pasture, but by 1930 only 145,000 acres of grass remained. Nearly all of the land level enough for tractor plowing had been tilled for wheat. One Meade County, Kansas, farmer wrote, "My tractor roared day and night, and I was turning eighty acres every twenty-four hours, only stopping for servicing once every six hours." A hired man drove the tractor from six o'clock in the morning to six o'clock at night, and the owner drove the remaining twelve hours.[16]

The same situation existed in other Kansas Dust Bowl counties. From 1910 to 1915, Morton County farmers planted an average of 1,177 acres of wheat. During the period from 1930 to 1935, however, 96,454 acres were planted annually. In Grant County, the increase for those same periods were from 1,509 acres to 140,626 acres respectively. During the same period in Hamilton County, wheat acreage rose from 1,459 acres to 85,440 acres, from 39,250 acres to 144,762 acres in Seward County, from 502 acres to 152,353 acres in Stanton County, and from 10,813 acres to 129,178 acres in Stevens County. During years of adequate rainfall, almost all of these acres were harvested, but in dry years nearly complete abandonment prevailed.[17]

Similar expansion occurred in the Texas Panhandle, where farmers planted 82,138 acres to wheat in 1909 and 582,827 acres ten years later. By 1929, Panhandle farmers planted 1,959,210 acres in wheat, or 43 percent of the area's cropland. By the great crash of 1929, twenty-six Panhandle counties produced nearly 66 percent of the Texas wheat crop. Much of this southern Plains expansion went relatively unnoticed. Had a similar amount of forest been cut down as grass plowed under, the threat to soil erosion would have been obvious. In the southern Plains, though, once the sod was gone, the land still looked much the same as giant wheat fields trailed off to the distant horizon. In the spring and autumn the land was green with wheat. Only later, when a severe drought developed, was the damage noticed.[18]

Certainly, the increased use of tractors and combines enabled southern Plains farmers to plant and harvest more wheat acreage than previously. As tractor sizes grew, farmers pulled larger plows. Combines also got bigger. In 1917, most combines had a nine-foot swath; three years later the average cut was twelve feet. By 1928 combines had sixteen or twenty feet cutter bars and could easily harvest 500 to 700 acres during the season. One newspaper reporter in the Oklahoma Panhandle described that harvest in the following manner:

> Occasionally we would run across a field where the old binders were being used, and the great number of shocks would attest the splendid production and the belief of some of the farmers that the old shocked and threshed wheat pans out best . . . . But when one sees a combine and tractor manned by one person sitting in the shade of a large umbrella, cutting a swath of wheat twenty feet wide, and not shaking down so much of the grain as the old ten-foot harvester did, and the clean grain falling into the wagon bed alongside the combine, as compared with the header driver, four barge men and two stackers not to mention the threshing crew, which are obviated by the wonderful new machine, one realizes its advantage, and the labor expense saved. A marvelous number of new auto trucks swept by us, and they carried loads both to and from the market. Going in they were loaded with wheat, and coming back they were loaded with new disc plows and machinery to be used in following the combines and headers in preparing the land for next year's crop, and were trailed in many instances with new drills to plant the coming crops.[19]

Southern Plains farmers got power machinery at a frantic pace. In a twenty-seven-county area, the value of new equipment rose from $16.1 million in 1925 to $36.0 million in 1930. However, the increased use of technology dictated further expansion. Tractors and combines were generally purchased on credit, and farmers broke more sod and planted more wheat to pay for them. In addition to the technological stimulus and relatively high yields (averaging nine to eleven bushels per acre), stable prices also contributed to expansion. Although wheat prices dropped more than a dollar per bushel during the postwar recession in 1921, the price averaged about a dollar per bushel for the next decade and provided a sufficient return to make continued expansion possible if not as profitable as many farmers wished. All of these factors then—increased use of technology, high yields, and stable prices—contributed to the breaking of more sod. By 1930, most of the good farm land had been planted. In southwestern Kansas, over 50 percent of the range had been plowed for cropland, and in Baca County, Colorado, about 60 percent of the sod had been broken for wheat by 1931.[20]

As wheat acreage increased, farm sizes also grew. From 1925 to 1930, the average farm size enlarged from 780 acres to 812 acres in the twenty-seven-county survey area. As farm size grew, the total acreage in farm land also expanded to more than 73 percent of the total land area, and crop acreage increased from 210 acres to 292 acres per farm. By 1930, 64 percent of all Dust Bowl farmers relied on cash grain crops for their livelihood. More cropland meant less protective grass cover. In 1920, 5.3 million acres in southwestern Kansas were in grass; by 1930, that figure had been reduced to 4.3 million acres. As the pastures declined in area, cattle overgrazed the remaining grass. The average value of farms increased from $12,325 in 1925 to $16,311 in 1930. More expensive property and machinery costs encouraged the growth of farm tenancy from 25.6 percent in 1920 to 38.5 percent in 1930. Tenants were often less concerned about soil conservation than were owners, and "suitcase" farmers made matters worse in some localities.[21]

Technological advance, high prices, and adequate precipitation encouraged agricultural expansion by absentee or "suitcase" farmers. Since wheat planting and harvesting takes only about six weeks a year, individuals living several hundred miles from their fields could engage in wheat farming provided they had the necessary access and equipment to do the work quickly. Many suitcase farmers had their own tractors, and automobiles provided convenient transportation to their lands. A person with two sources of income could afford to gamble on wheat production, and such risks were often taken in the northern half of the Dust Bowl. There, vacant grasslands bordering well-developed central Kansas wheat lands lured the suitcase farmers. In 1921, a Denver-based suitcase farmer broke 32,000 acres in Greeley County, Kansas. Three years later he had 50,000 acres in wheat. Not all suitcase farmers operated on such a large scale, but they did have one similar characteristic—flexibility. If a crop failed, a suitcase farmer still had another income, and his livelihood did not depend upon him remaining on the land. If a wheat crop did not look profitable, a suitcase farmer could abandon his fields to the mercy of the wind. Drought and blowing dust encouraged such cutbacks. When suitcase farmers abandoned their land, they seldom returned to apply the proper soil conservation techniques to keep it under control.[22]

Still, suitcase farmers were not much different, at first, from other southern Plains farmers. Most farmers in the region gave little thought to plowing crop residues under to maintain or increase soil humus. Mechanical difficulties as well as labor and equipment shortages and costs often prevented them from working stubble into the ground. The common practice of burning off the stubble hindered the return of crop residue to the soil, and as its organic content decreased, the soil became less productive. Farmers also hesitated to work wheat stubble into the soil because it caused a temporary fertility depression. During a period of drought, though, plowing under the stubble more than offset the fertility loss because it increased the soil's ability to absorb moisture. Farmers all too often allowed their herds to graze in the stalk fields until they had consumed every bit of vegetation, thus leaving the ground unprotected from the wind. Finally, continued cultivation pulverized the soil so that by five to seven years after the sod had been broken, it was in an excellent condition for blowing. After a farmer broke the native sod, the productivity of the soil steadily decreased because of continual loss of organic matter, and the loss of organic material made the soil more susceptible to wind erosion since it was then less able to absorb moisture and thus less supportive of vegetative growth. Continued plowing and overgrazing made the soil almost immediately subject to wind erosion during a drought. Sandy soil could be completely blown away to the depth of the plowing within a few days.[23]

Drought was also a problem that contributed to the creation of the Dust Bowl. Ordinarily, the southern Plains region receives approximately eighteen inches of annual precipitation. This amount is adequate for a satisfactory crop yield only when it is carefully conserved; however, southern Plains farmers seldom tried to conserve the moisture in the subsoil prior to 1935, so the drought of the 1930s proved devastating. The Springfield, Colorado, area averaged more than three inches below normal precipitation between 1931 and 1935; the Meade, Kansas area had a five inch annual deficit from 1931 to 1933; the Goodwell, Oklahoma, area was down nine inches in 1932 and 1933; and, the Amarillo, Texas, area averaged more than seven inches below normal from 1933 to 1935. In this region where crops need every drop of moisture that falls, a deficiency of even a few inches can mean the difference between a bountiful harvest and economic disaster. Moreover, winter snows were insufficient to protect the soil, and winter contributed to erosion by loosening the ground with alternate freezing and thawing. The lack of precipitation and the harsh winters thus aided the prevailing winds of the southern Plains, increased evaporation, weathered the soil, and brought dust storms to the southern Great Plains.[24]

Still, the wind erosion problem of the southern Great Plains did not occur because farmers grew too much wheat, but because the drought prevented them from growing hardly any wheat at all from 1932 to 1940. During years of normal precipitation, the extensive root system of the wheat plants held the soil and offered excellent protection against wind erosion. In the droughty thirties, however, the inadequate moisture supply prevented a suitable growth of ground cover in the early spring "blow season" of February, March, and April. The drought then began a chain of events, the first of which was crop failure. Abandonment of land without a protective soil cover in turn allowed the nearly constant winds to begin erosion. The dust storms that followed drifted the loose soil, ruined additional land, and contributed to more crop failure. Wind erosion worsened during 1933, when an insufficient and poorly distributed rainfall together with winds of above average velocity brought widespread damage to the southern Great Plains. By August 10, Goodwell, Oklahoma had experienced more than thirty dust storms. In May 1934 the drought was the most severe on record, and the erosion problem steadily worsened as the wind stripped the top soil to the depth of the plowing in many parts of the Great Plains.[25]

In retrospect, many factors contributed to the creation of the Dust Bowl—soils subject to wind erosion, drought which killed the soil-holding vegetation, the incessant wind, and technological improvements which facilitated the rapid breaking of the native sod. The nature of southern Plains soils and the periodic influence of drought could not be changed, but the technological abuse of the land could have been stopped. This is not to say that mechanized agriculture irreparably damaged the land—it did not. New and improved implements such as tractors, one-way disk plows, grain drills, and combines reduced plowing, planting, and harvesting costs and increased agricultural productivity. However, the new technology also had negative effects. Increased productivity caused prices to fall, and farmers compensated by breaking more sod for wheat. At the same time, farmers gave little thought to using their new technology in ways that would conserve the soil.[26]

In 1931, a bumper wheat crop accompanied by drought brought economic disaster to the southern Great Plains. In the Dust Bowl states, the price of wheat fell from an average of $.99 per bushel in 1929 to $.34 in 1931. As the price of wheat plummeted, good farmers could no longer afford to practice soil conservation techniques such as listing, terracing, or strip cropping. Even minimal soil exposure began to contribute to soil blowing, and some of the worst dust storms came from areas where less than half of the acreage had been planted in wheat. Although a wind of thirty miles per hour was often necessary to start soil movement on the best lands, once it began to blow, a wind velocity of less than half that figure could easily stir the soil into the air.[27]

As the twin problems of drought and soil blowing became worse, conservationists, soil scientists, and southern Plains residents began to classify dust storms. H. T. Coleman, a meteorologist at the Amarillo, Texas, Weather Bureau, claimed there were two classes of dust storms—southwestern and northern. He believed that the eastward movement of a low pressure center triggered the southwesterly dust storms. These storms lifted the soil to great heights, often over ten thousand feet, but the surface density was not great. The northerly storms were much worse and were commonly known as "black blizzards." In these storms, the dust boiled or rolled along the surface and reduced visibility to zero. The Kansas Academy of Science identified three classes of dust storms—rectilinear, rotational, and ebullitional. Within these three classes, seven "species" of dust storms were recognized. These storms ranged from the relatively trivial "sand-blow" type to the "funnel storms" which could transport dust three to four miles in altitude and carry it two thousand to three thousand miles away.[28]

No matter how dust storms were classified, they were all disagreeable. Terrible though many of them were, the power, intensity, and destructiveness of the storms was somehow fascinating. When the storms began, most Dust Bowl residents had to admit they had never experienced anything quite like them before.

CHAPTER THREE

The Storms:
1932-1940

The storms in the southern Plains began in late January 1932, when a duster swept across the Texas Panhandle. This gale struck Amarillo from the southwest shortly after the noon hour on January 21 with sixty-mile-per-hour winds and a dirt cloud that reached ten thousand feet high. Almost blinding in its intensity, the storm caused minor damage to plate glass windows before it moved into Oklahoma and Kansas where it also damaged property and destroyed wheat fields. The Amarillo weather bureau called this storm "most spectacular" and "awe-inspiring." On February 2, near Lubbock, Texas, the wind rose with the sun and by noon a "very nice" sandstorm was blowing. Similar dust storms throughout the spring darkened the sky, restricted travel, and prompted old timers to call them the worst ever. Some of the lighter soils blew severely at this time, but most dust storms in 1932 were local and were largely confined to sandy lands where wheat, corn, or cotton crop failures had left the ground bare.[1]

Dust storms became more common during the following spring, when insufficient and poorly distributed rainfall, together with winds of above average velocity, brought widespread damage to the southern Great Plains. Lights of automobiles, stores, and homes had to be turned on in the afternoon. Sunlight took on a violet-greenish hue, cars became stranded along highway ditches, and cattle huddled against the dust like they would against wind-driven snow. In mid-May, Charles A. Lindbergh and his wife were flying from Albuquerque to Kansas City when a dust storm forced them to make an emergency landing on the plains northwest of Canadian, Texas. The Lindberghs spent a grimy night in the plane before continuing their flight the next morning. Still, bad as the dusters were becoming, most farmers were more concerned with crop failure, low prices, and the depression than they were with wind erosion until a dust storm in mid-November 1933 swept beyond the Plains and deposited soil as far east as Lake Superior.[2]

That dust storm covered a vast expanse of the central United States from Montana to the Great Lakes and southward to the lower Mississippi Valley—a region larger than the combined area of France, Italy, and Hungary. In South Dakota, where drought plus crop and pasture failure accentuated the storm, old-timers again considered it the worst in their memory. Not until the spring of 1934, however, did plainsmen and the entire nation become seriously alarmed about the growing menace of dust. In mid-March a dust storm uprooted wheat in Meade County, Kansas and "did about everything but take the last hope of the people." By April, powdery dirt was drifting over fences and the wind had exposed the subsoil in some places. On April 12, dust reduced visibility to less than one mile in Baton Rouge, Louisiana. Drought conditions became worse, and in May the wind stripped the soil to the depth of the plow in parts of the Great Plains, removed approximately 300 million tons of soil, and deposited it over the eastern half of the nation including Washington, D.C., and New York City; dust from that storm even fell on ships five hundred miles out at sea. For the first time, many easterners smelled, breathed, and tasted soil that came from the Great Plains. Few liked the experience.[3]

Twenty-one inches of snow containing more than two inches of moisture fell over the Texas Panhandle in March 1934, causing many local residents to believe that the drought had ended. But this snowfall was not general across the Dust Bowl, and it did not contain enough moisture to prevent further soil blowing. The following month in southeastern Colorado, Baca County experienced a duster that struck like a "black tidal wave" with winds of fifty miles per hour. A cloud of fine, powdered earth obscured the sun for the next six days and forced residents to remain indoors with wet towels muffling their faces. More dust storms followed.[4]

Light rains fell across portions of the Dust Bowl in April and May, but the storms continued. S. D. Flora, a federal meteorologist, reported that Kansas received less moisture in April than during any comparable month for the previous twenty-four years. As a result, the powder dry soil was at the mercy of the winds. Blowing soil became more frequent and widespread than at any other time since weather records had been kept for the Dust Bowl area. The majority of the western counties in Kansas received less than one-half inch of precipitation during April and several extreme southwestern counties failed to get as much as one-quarter inch. Kansas had been drier during the first four months of the year on only two previous occasions—1893 and 1910. For the state as a whole, twenty of the twenty-eight previous months had been deficient in rainfall, and some localities in the Dust Bowl were even drier.[5]

In early May, a dust cloud extended fifteen hundred miles from the Rockies to the Great Lakes, and nine hundred miles from the Canadian border to Oklahoma. This storm lasted thirty-six hours and reduced visibility to less than a mile in Kansas City, Chicago, St. Louis, Des Moines, and St. Paul. It so dramatized the drought and wind erosion damage to crops that wheat leaped the five cent limit on the Chicago Board of Trade, and crop experts predicted a dollar a bushel quotation within a week.[6]

From early June 1934 through February 1935, dusty conditions prevailed in local areas across the central and southern Plains. Although many of the dust storms of 1934 were violent, nothing quite equalled the vicious black blizzards that blew during 1935. On February 21, one of the most severe dusters in twenty years struck western Kansas from the north and rolled southward into Oklahoma and Texas. The dust boiled up from the tinder dry fields, downed telephone lines, and brought traffic to a standstill. At 12:30 P.M., day became night in Colby, Kansas, and rural schools closed when the dust got so thick that classwork was impaired. During the height of the storm, a western Kansas farmer drove his car off the road and began to walk the two miles home. Searchers found him the next day suffocated after having walked half way home. The storm struck Amarillo with gale force winds of fifty miles per hour. Although visibility was limited to 1/4 mile, airlines maintained regular schedules. By now, many Dust Bowl residents were getting accustomed to the storms, and, while travel was often uncomfortable, it was not impossible for people who did not mind getting dirty. At Alva, Oklahoma, the storm sifted dust into the college gymnasium to such a degree that a basketball game in progress had to be stopped. After the storm, housewives turned garden hoses on their windows before they could give them a regular cleaning. Merchants reported a brisk sale of brooms, mops, brushes, and vacuum cleaners; dry cleaning businesses also enjoyed an increase in business. Several days later, another dust storm halted all air, bus, train, and automobile travel.[7]

During March and April 1935, Amarillo, Texas, and Dodge City, Kansas, had twenty-eight and twenty-six dust-laden days respectively. The month of March was particularly dirty in eastern Colorado, where dust storms occurred frequently from March 12 to March 25 causing six deaths and more than a hundred serious illnesses. There the dust formed drifts like snow from a few inches to more than six feet deep; livestock in considerable numbers died from ingesting too much dust covering their feed and grass; schools also closed.[8]

In early March 1935, a black blizzard turned day into night in the Texas Panhandle; it broke glass windows, damaged railroad communication, and caused a woman to faint in the lobby of a downtown Amarillo hotel due to difficulty in breathing. It was the worst dust storm in thirty-seven years for one farmer near Lubbock, Texas. A Garden City, Kansas housewife reflecting on a mid-March duster said, "All we could do was just sit in our dusty chairs and gaze at each other through the fog that filled the room and watch the fog settle slowly and silently, covering everything." Although the doors and windows were shut tightly, the dust still sifted into every crevice, including cupboards and closets. Following the storm, she continued, "Our faces were as dirty as if we had rolled in the dirt; our hair was gray and stiff and we ground dirt between our teeth." In Wichita the department of geology at the university "weighed" the atmosphere and estimated that 5 million tons of dust hung one mile thick over the thirty square miles of city—about 167 tons per square mile. At Alva, Oklahoma, city police blocked the highways and prohibited motorists from leaving town. In Clovis, New Mexico, the secretary of the chamber of commerce noted that on many days he could barely see the nine-story hotel across the street from his office. "Business dies," he wrote, "on those days."[9]

By mid-March, the dust storms had become so commonplace in western Kansas that residents began to accept them as a basic aspect of daily life—even going to bed in dust-laden air only to shake off the covers in the morning. Often the mere washing of hands brought an earthy odor similar to that which comes from a spring shower on a dusty road. One defiant Dust Bowl resident wrote on the window of a defunct High Plains bank: "Ashes to ashes and dust to dust, the men folk raved and the wommin cussed, take it and like it, in God we trust." Take it they would, but they did not like it. Little did they know that the dust storms would get worse. On March 26, near Dighton, Kansas, a freight train rammed a special passenger train of Civilian Conservation Corps recruits as it backed onto a siding. The dust was blowing so badly that the engineer could not see the warning signal. Thirty recruits were injured, and two trainmen were seriously hurt. By now the windows at the Dodge City hospital were plastered shut to keep out the dust. Farther to the north, at Hays, Kansas, on the fringe of the Dust Bowl, hospital nurses placed wet cloths over the faces of their patients. Wildlife suffered too, and the storms killed jackrabbits by the thousands. By the end of March, the dust was so bad in Dallas, Texas, that numerous people wore cloth pads over their noses and mouths.[10]

Four days before the notorious black blizzard of April 14, 1935, a severe dust storm rolled across the Texas and Oklahoma Panhandles and into Kansas. For more than twenty-four hours it paralyzed traffic and set a record for the intensity and duration of a daytime storm. In Boise City, Oklahoma, forty-mile-per-hour winds accompanied the dust, and concerned local businessmen escorted school students home in the early afternoon. At nearby Guymon, the storm lasted forty-eight hours. In the heart of the Texas Panhandle at Amarillo, airport officials estimated that the dust cloud reached fifteen thousand feet.[11]

That storm struck Garden City, Kansas, before 2:00 A.M. and raged into the day. By noon, it was still black as night, and all business and social activity ground to a halt. Rain fell through the dust cloud in the afternoon, bringing mud balls from the sky. In Baca County, Colorado, every school closed because the dust had been blowing badly there for five days. Local stores exhausted their supplies of dust masks and sponges. Visibility was so poor that one truck driver spent five and a half hours driving from Springfield to Lamar—a distance of fifty miles. Near Deerfield, Kansas, trains barely crept along while crewmen walked ahead to determine whether drifted dust was piled high enough on the tracks to derail the engine or cars. Handcar crews patrolling the Santa Fe line with shovels were unable to keep the tracks open because the dust drifted over the rails faster than they could scoop it off. Trains were stalled at Dodge City and Syracuse, Kansas, and another derailed farther east near Great Bend. Visibility was so bad that a wrecking crew sent to rescue the train could not work and was itself stranded for two days.

In the Texas Panhandle, the Fort Worth and Denver Railroad service between Dalhart and Texline was interrupted until snow plows removed the drifting sand from the tracks. Rural schools now began to close a month early because board members were tired of shoveling dirt out of classrooms before the daily lessons could begin and because it was dangerous for pupils to leave school for home during a duster.[12]

As the soil sifted out of the Dust Bowl, people as far east as Memphis covered their faces with handkerchiefs. Texas state senators put on surgical masks while the legislature was in session to make breathing easier and more healthful. Senator Ben G. O'Neal of Wichita Falls protested, however, calling out: "Point of order, the governor is trying to gag the Senate." When April ended there were only five dust-free days for the month in southwestern Kansas; on fourteen days, visibility was less than one quarter mile. April had also been the windiest month on record, and the average velocity of 16.2 miles per hour greatly intensified the erosion problem. At Dodge City, April was also the second driest on record with only .03 inch of rain. Light rains in early May brought renewed hope that the drought had ended. One-half inch of rain fell at Springfield in Baca County; this was the first moisture since January and the best rain since 1932. One Dust Bowler remarked, "A man could walk to church and arrive with his shirt still clean for the first time in many a Sunday." This optimism was premature, and the dust storms continued.[13]

In August 1935, after the ceilings of two Dodge city homes collapsed from the weight of accumulated dust in the attic, one enterprising man obtained an oversized carpet sweeper and went into the attic cleaning business. The cleaner blew the dust into gallon pails which were then emptied outside. By late summer he had cleaned 227 southwestern Kansas homes, taking an estimated two tons of dust from each attic.[14]

Dust storms with high winds were common over the southern Plains during the late winter of 1935 and early spring of 1936. On March 1 a light dust storm began in east-central

Colorado and by the fourth had moved into the Oklahoma and Texas Panhandles. Six days later a moderate to heavy dust storm occurred in this area. A dust cloud enveloped the entire region of Colorado east of the 104th meridian on March 11. Dust was still blowing on March 13, and nearly the entire area immediately east of the Rockies had visibility reduced to less than one mile. With a wind of thirty-three miles per hour, the sand and dust removed paint from automobile bodies and pitted windshields. As this storm moved into Oklahoma, the dust clouds turned from dark black to a yellowish-red color and restricted visibility to twenty-five feet in the panhandle.[15]

Through the remainder of March the dust storms continued, and kept Baca County in semidarkness from three to twenty-four hours a day. Black blizzards raging in southeastern Colorado and southwestern Kansas on March 25 and 26 were reported the most severe and longest in duration of any before. Dust storms were becoming worse, in part because the coarse, granular structure of the soil particles was breaking down due to the drought and the constant blowing and shifting of the soil. Much of the topsoil had become a fine powder which winds of relatively low velocity could pick up and carry farther than in the past. With the southern Plains becoming more desertlike with each passing month, one resident claimed the only difference between the Dust Bowl and the Sahara Desert was that a lot of "damned fools" were not trying to farm the sands of northern Africa.[16]

The heart of the Dust Bowl, which suffered the most severe wind erosion, was subjected to more dusty days in 1937 than during the previous year. Goodwell, Oklahoma experienced 134 dirty days from January 1 to September 30. When the frost left the ground, the soil was in perfect condition for blowing because the alternate freezing and thawing of the ground had pulverized it even further. Southeastern Colorado was particularly dusty where light to moderate winds blew the soil badly. In Baca and Prowers counties, seven dust storms with an average duration of six hours reduced visibility to three hundred yards. New Mexico also experienced considerable soil erosion in January, and by the end of the month soil was beginning to blow in southwestern Kansas.[17]

The storms became worse in February and carried soil beyond the Dust Bowl. Almost all of the top soil was by now dry to a depth of several feet and the wind easily drifted soil many feet deep. Deserted farm homes became common across the region. Kansas experienced vigorous storms during the month, and the Oklahoma Panhandle had twenty days of heavy dust fall. One dust storm reduced visibility in western Kansas and the Texas and Oklahoma Panhandles to five feet for most of an entire day. In the Oklahoma Panhandle, farmers rode their tractors with goggles and handkerchiefs over their faces. But farmers were not the only ones who suffered the effects of soil erosion. Townsmen did too, and the representatives of the Guymon chamber of commerce donned dust masks for their annual membership drive.[18]

The dust storms became worse in March and April 1937. The most severe storms occurred during the third week of March and the prevailing westerlies carried the soil to the Middle Atlantic and eastern Gulf states. During the height of the storms, visibility was often reduced to zero. Flying particles of sand lacerated the wheat crop, and the pastures became so covered with dust that grass was unfit for cattle to eat. In early March one Associated Press reporter wrote: "It is strange that none of the bright *young* minds in Washington, D.C. has suggested that Congress pass a law to prohibit dust storms." He argued that since Congress, under its authority to regulate commerce, had the power to control and regulate production, it could declare dust storms illegal since they interfered with productivity. He recognized, however, that controlling the weather is a difficult undertaking. But, for him, it was no more foolish than some other things which the Roosevelt administration had attempted.[19]

Widespread rain and snow in late March temporarily alleviated the dusty conditions across the southern Great Plains. However, strong winds in April brought dust storms to some sections of New Mexico every day. Kansas had nineteen dust-laden days during the month, ten of which were especially severe. In eastern Colorado, a dust storm late in the month made airplane radio equipment useless, and many planes made forced landings. One such landing occurred at Sterling, where one thousand automobiles and some fire trucks lighted the airport with headlights so that a transport plane could see the runway through the dust.[20]

During May 1937 dust storms again plagued the southern Plains, damaging the wheat crop and injuring livestock. The storms were the most frequent from May 18–28, and portions of Oklahoma suffered twenty-one days of severe dust. Southwestern Kansas reported up to fifteen such days. Goodwell, Oklahoma was particularly dusty during May. The report of the weather bureau is illustrative:

> At Goodwell the visibility was reduced to 1/2 mile on the 1st, 150 yards on the 3d, 1/2 mile on the 4th, 1/4 mile on the 6th, practically zero at 7:45 P.M. on the 7th, 1/2 mile on the 10th, 1/4 mile on the 11th, 100 feet at 5:45 p.m. on the 16th and at 6 p.m. on the 17th, 150 feet on the 19th at 6 p.m., 1/2 mile on the 20th, zero on the 21st from 7:09 p.m. to 11:09 p.m., 1/4 mile on the 23rd, 100 feet at 8:45 a.m. on the 24th, 1/2 mile on the 25th, 1 mile on the 26th, only 10 feet for a short time at 3:45 p.m., with some rain falling while dust was heaviest, and 1/2 mile on the 30th.

The worst dust storm since April 14, 1935, struck Boise City, Oklahoma, on May 22, 1937. Absolute darkness lasted fifteen minutes. At Elkhart, Kansas, high school commencement exercises were postponed because a dust storm began a few hours before the ceremony was to begin. Numerous storms also occurred in southeastern Colorado at this time, where black blizzards made the days "dark as night" and "absolutely black" according to reports. In late May, a dirt cloud covered the entire area east of the 103rd meridian, reducing visibility to as little as twenty feet for from four to seven hours. Similar conditions prevailed over other portions of the Dust Bowl.[21]

In June 1937, scattered rainfall reduced the number of dust storms, but the storms continued periodically for the remainder of the year. In fact, the densest dust storms occurred relatively soon after rains, which in some areas exceeded three inches. For the most part, though, the dust storms were not as widespread as during March, April, and May. No storms were reported in the eastern two-thirds of Kansas during June 1937. The Oklahoma Panhandle had only twelve dusty days. Dust continued to blow occasionally in July, although storms were generally reported as being "light to moderate in character." The frequency of dusty days during July varied from one in portions of Texas and New Mexico to nine in Kansas, ten in southeastern Colorado, and seventeen in Oklahoma. Similar conditions existed through August. September rains helped hold the soil, although infrequent but occasionally dense dust storms occurred.[22]

No dense dust storms struck in October but heavy dust sometimes fell in portions of the southern Plains. At Amarillo, fifty-four mile-per-hour winds with blowing dust reduced visibility to one-half mile for two hours. Dust fell across the Oklahoma Panhandle on October 1, 4, 12, and 18. On the latter day, visibility at Goodwell, Oklahoma, was reduced to one-half mile from 11:00 A.M. to 4:00 P.M. Southwestern Kansas also experienced local storms, and dusty conditions not severe enough for classification as dust storms were present in many southwestern counties on October 3, 4, 6, 12, and 18. The situation was much the same in southeastern Colorado.[23]

Precipitation was subnormal most of the autumn. Kansas received only 51 percent of its average for November. The soil was drier than usual for that time of year, and conditions for blowing were perhaps greater than at any time since the dry cycle began. Still, only a few dust storms were serious enough to be classified as dense. Light to moderate dust storms occurred in the Oklahoma Panhandle on seven days in

November. Several moderate storms swept across western
Kansas, but only a November 12 storm was dense enough to
stop the automatic sunshine recorder of the Dodge City
Weather Bureau. Light dust fell over Baca County on numerous days, but compared to the severity of past dust storms, it
went almost unnoticed. December remained subnormally
dry, but no heavy dust storms were reported by the weather
stations, largely because the wind movement was lower than
in previous years.[24]

The dust storms of 1938 were far less extensive across the
Great Plains. In the Dust Bowl, however, occasional storms
were as severe as those of the preceding years. Continuation
of the drought and dry subsoil conditions made dust storms
numerous during January and damaged the winter wheat
crop considerably. In western Oklahoma, the storms blew
seed out of the furrow, and dust also plagued eastern New
Mexico. During the latter half of the month, dust was almost
continually in the air throughout the southern Plains. On January 18, 1938, near Sublette, Kansas, a train of the Dodge
City and Cimarron Valley branch line was stalled for eighteen
hours because of dust drifts covering the tracks. A combination snow and dust plow cleared the three to six feet drifts
(one of which was fourteen hundred feet long). In southeastern Colorado, one dust storm continued for sixty hours, and
reduced visibility to less than a hundred yards through Baca
and Prowers counties.[25]

February rains and snows fell over most of the Dust Bowl
in 1938, settled the dust, and helped to retard soil blowing.
However, eastern Colorado did not receive this precipitation,
and blowing dust caused considerable damage to the wheat
crop in Las Animas County. Precipitation which fell was not
sufficient to restore the deficient subsoil moisture, and
before long the soil began to move again across the Dust
Bowl. In March 1938, more dust storms occurred than in any
other month for the year, and they were most serious at the
end of the month. Hardly a day passed in Colorado when the
dust did not blow; heavy dust storms even occurred within an

hour or two following a heavy rain. On March 24, dust closed the Pueblo airport. In Dodge City, light dust fell on thirteen days, thick dust on nine days, and dense dust with visibility less than one-quarter mile on three days. During the month the Santa Fe Railroad had to plow dust from the tracks twice a week between Dodge City and Boise City, Oklahoma. Here, the dust drifted five to six feet deep. Twelve hours after Dodge City received .75 inches of rain, dust reduced visibility to less than one-half mile.[26]

The Dust Bowl received subnormal rainfall in April 1938, but nearly all of the severe dust storms occurred before the fifteenth. Light dust fell on occasion during the latter half of the month. In Colorado, however, black blizzards "squelched" the southeastern counties; at times the dust was as intensive and destructive as any storm during the worst year, 1935. Wind blown dust lacerated wheat plants or completely covered them. Sand dunes reached five feet high in southern Lincoln County, and flying sand pitted automobile windshields. At Amarillo, the worst dust storm of 1938 came with an early April blizzard. Sixty-mile-per-hour winds with gusts up to eighty miles per hour drifted a mixture of dust and snow.[27]

Above normal amounts of precipitation fell over the Dust Bowl in May 1938 except for Texas and New Mexico. Kansas received over twice the monthly average of rainfall. Even so, unusually hard winds in mid-May created dust storms in southwestern Kansas and the Oklahoma Panhandle. However, above average precipitation and low seasonal wind velocities gave southeastern Colorado much needed relief from destructive dust storms. Vegetation began to sprout and hold the soil. Nevertheless, soil blew during ten days in May 1938 in this section of the Dust Bowl; two of those storms were serious.[28]

More rain fell in portions of the Dust Bowl in June, but eight storms reduced visibility from one hundred feet to zero for up to eighteen hours in Baca County. As a whole, these storms were less extensive and intensive because large sections of southeastern Colorado were beginning to "heal over" with soil holding weeds. Still, large tracts of land remained bare. There, the dust blew with the slightest wind movement. On June 10 automobile windshields were sand-blasted near Dodge City to such an extent that new glass was required on some cars. At this time, the worst wind erosion area was southwestern Kansas and the Oklahoma Panhandle. The Santa Fe Railroad was still using snow plows on a regular basis to remove drifts over a portion of the line. One drift extended 2,500 feet and averaged from two to seven feet deep.[29]

The Dust Bowl received above average precipitation in July 1938, and for the first month since January 1937 no dust storms struck Dodge City. Only light dust and widely scattered storms occurred throughout the remainder of the summer. Additional rains and light winds prevented further serious storms for the last half of the year.[30]

The dust storms of 1939 were less widespread and less intense than at any time since 1932. Although the drought prevailed, only light storms developed across the southern Plains. In fact, the Dust Bowl had only five days of bad soil blowing during January 1939. Occasionally, dust reduced visibility to zero for several hours in Baca County, damaged wheat fields, and blew thin stands of sorghum out of the ground. Storms such as these swept across the Oklahoma Panhandle on January 4, 27, and 31.[31]

On February 1, 1939, H. H. Finnell, regional conservator for the Soil Conservation Service, indicated that subsoil moisture had reached a depth of forty-eight inches in parts of the Dust Bowl. Only 9.5 million acres were still subject to severe wind erosion, down from 50 million acres in the winter of 1935–1936. The Amarillo office of the Soil Conservation Service optimistically reported that the Dust Bowl had shrunk to one-fifth its previous size and that the soil was in its best condition since 1932.[32]

Still, dust storms increased in frequency and intensity in March 1939, and in April southwestern Kansas had sixteen

dusty days. Heavy dust storms swept across the Oklahoma
Panhandle on April 10, 17, 18, 20, and 30. Southeastern Colorado had considerable dust on nine days; these storms lasted
an average of seven hours with visibility from 0 to 200 feet on
the leeward side of blowing fields. Periodic dust storms ranging from light to heavy in intensity occurred throughout the
remainder of the spring. In June 1939 southwestern Kansas
experienced dust storms during almost half the month. In
Colorado, dust storms were more frequent and damaging
than at any other time that year. During the remainder of the
summer, local but sometimes heavy dust storms struck. Only
a few storms were reported during the autumn and early winter. Although visibility was often markedly reduced, these
storms were light in character. By the end of the year, the
Dust Bowl was confined to the extreme sections of southeastern Colorado and southwestern Kansas.[33]

Throughout most of the 1930s, then, continued drought
and crop failure allowed the soil blowing. The number of dust
storms increased in most parts of the Dust Bowl from 1934 to
1938. The amount of acreage subject to wind erosion also
expanded during that time in spite of the increasing efforts of
farmers to bring their fields under control by various soil and
water conservation methods. In Oklahoma, some agriculturists warned that the Dust Bowl was creeping eastward at the
rate of thirty miles each year. Faced with this threat some
people speculated that in time Great Plains and east coast cities would be covered with dust like those of many ancient
civilizations. Others speculated that between the Alleghenies
and the Rockies a "forbidding desert" was in the making,
"similar to the sands that once knew the footprints of the
Queen of Sheba." The tenacious spirit of the Dust Bowlers is
a significant reason why such a desert did not develop.[34]

CHAPTER FOUR

Life in the Dust Bowl

The dust storms which began in 1932 and peaked in 1935 continued intermittently during the spring months of the next four years, but by 1940 a return of the wet cycle ended them. During that eight year period, though, most Dust Bowlers tolerated and endured the storms.

As the black blizzards darkened the sky, residents sealed windows with gummed tape or putty and hung wet sheets in front of living room and bedroom windows in futile attempts to filter the air. Some people spread sheets over their upholstered furniture and wedged rags under doors to keep the dirt out of their homes. Plates, cups, and glasses remained overturned on the table until the meal was served from the stove. Pans were quickly covered to keep out as much dust as possible. Housewives learned to keep liquids sealed in mason jars, and sometimes mixed bread dough in bureau drawers to escape the dust-laden air. In April 1935, fifty-three housewives in Guymon and Goodwell, Oklahoma, estimated the cost of cleaning house for a month at nearly $30.00, based on a labor rate of twenty-five cents an hour. Laundry and dry cleaning costs mounted. However, one Dust Bowl resident testified that the extra housecleaning was "just what the women's figures needed."[1]

Surgeons and dentists fought the problems of sterilization. Railroad engineers sometimes failed to see the stations and had to back up the trains. Electric lights even in daytime dimmed to a faint glow along streets. Motoring was particularly hazardous during a duster because of poor visibility and dust drifts across highways. Static electricity which accompanied the storms caused automobile ignition systems to fail; many cars stalled until a storm had passed. Even windmills, pump handles, and frying pans became so highly charged that a good shock might be delivered to anyone who touched them. Motorists learned to attach drag wires and chains to their cars to ground this electricity and prevent short circuits and stalling.

Some agriculturists blamed the static electricity for helping to kill the wheat sprouts and for preventing germination of seeds by killing the soil. One local editor claimed that the dust storms "generated enough free electrical current to do all the farm work in the Southwest if it could be properly captured and utilized." Even though these claims were exaggerated, the storms did create a good deal of static electricity which was a constant nuisance. Another journalist reported in all seriousness that the static electricity was sometimes so powerful that it electrocuted jackrabbits. He did admit, however, that since autopsies had not been conducted on any of these animals, their deaths might have been attributable to having eaten too much dust.[2]

Although one may question whether static electricity killed the rabbits, one thing was certain—the jackrabbit population had proliferated beyond imagination. Whether the rabbits increased in the Dust Bowl because the drought improved breeding conditions or they migrated into the area in search of grass no one has yet determined. In any event, hundreds of thousands of jackrabbits competed for the sparse grass available, and Dust Bowlers were forced to begin an extermination campaign. Almost every Sunday people gathered on selected farms outside their local communities to take part in a rabbit drive. Once assembled the crowd dispersed into a large circle that might be over a mile in diameter. As they walked to the center the rabbits ran before them.

No guns were allowed—only clubs. As the circle tightened the rabbits were driven into snowfence pens where they were beaten to death. These Sunday rabbit drives were announced in local newspapers where editors urged, "Everyone who has a club of any kind is invited to participate in the drive." One Dust Bowl newspaper called for a "grand and glorious" rabbit drive on April 14, 1935, "unless the dust is too terrific." As things turned out, it was.[3]

Becoming lost in a black blizzard was a traumatic and sometimes fatal experience. On March 15, 1935, a black blizzard struck Hays, Kansas, catching a seven-year-old boy away from home. The next morning a search party found him covered with dust and smothered. A hundred miles to the west, the same storm stranded a nine-year-old boy; a search party found him the next morning alive but tangled in barbed wire. Farmers overtaken in their fields by a duster sometimes found their way home by following a fence wire with their hands or by crawling on their knees along a furrow which led them to the end of the field and the protection of their trucks. Once, when a local movie ended, a theater-goer crawled along the curbing until he reached his neighbor's house. There, the porch light looked like a candle enveloped in dense smoke.[4]

The dust storms brought health hazards. Laboratory experiments revealed that the dust contained a high silica content which had a poisoning effect in the body similar to that of lead. It weakened one's resistance to disease and became exceedingly irritating to the mucous membranes of the respiratory system. Although the dust did not contain any disease-carrying organisms, it did contribute to acute respiratory infections such as sinusitis, pharyngitis, laryngitis, and bronchitis, particularly for the very young or very old. It also increased the number of deaths from pneumonia. Dust in the lungs has a similar effect to that of silicosis which is common to quarry miners. The fine particles literally cut into the lung tissue. Illnesses characterized by respiratory irritation and choking were commonly referred to as "dust pneumonia."

Doctors treated it in the same manner as ordinary pneumonia. In order to treat the growing number of respiratory cases,
the Red Cross opened six emergency hospitals in Colorado,
Kansas, and Texas during the spring of 1935 and issued a call
from its "dust headquarters" in Liberal, Kansas, for 10,000
dust masks for immediate distribution to the five-state Dust
Bowl area. Local residents of Guymon, Oklahoma, also
turned two church basements into emergency hospitals. Dr.
William DeKline, the national Red Cross medical director,
used the Red Cross to coordinate emergency health care
activities in the Dust Bowl. By the time the Red Cross left the
area on May 29, 1935, it had distributed 17,700 dust masks
and sent nurses to 1,631 homes, mostly for dust related
illnesses.[5]

In Colorado, the state board of health ordered an investiga-
tion to determine whether dust was merely a contributing factor or the specific cause of deaths. Governor Edwin C.
Johnson promised oxygen tents, nurses, and other medical
supplies for relief of respiratory illnesses. Children suffered
particularly from dust inhalation. Insufficient hospital facilities was a serious problem, and the six physicians in Lamar
were often unable to reach patients because of drifted roads
and highways. In March 1935 the Springfield morgue had six
bodies—all victims of dust pneumonia. One was a seventeen-
year-old girl, another a seventy-five-year-old man. More than
a hundred people were seriously ill in Baca County at that
time. Kenneth Welch, relief administrator for that county,
reported that dust pneumonia was rapidly increasing among
the children and that men by the score were sending their
families out of the county. In late April, an estimated three
hundred persons in Beaver County, Oklahoma, were suffering from respiratory illnesses. Three people had died there
from dust pneumonia; nine had died in nearby Liberal.[6]

In Ford County, Kansas, the national Red Cross granted
money for dustproofing homes. Twenty carpenters employed
by the Kansas Emergency Relief Committee were to caulk
windows and seal them with glass cloth. One room in each

demonstration home was sealed in this manner; after that had been done the owner could seal the remainder of the house himself, or have relief labor do it. Selection of demonstration homes was based on need and family illness.[7]

The health situation worsened when Kansas suffered a measles epidemic. From January to early June 1935, 40,000 measles cases were reported, and in the first four months there were 145 deaths. Still, health officials were primarily concerned about respiratory infections. One Dust Bowl hospital reported that 12 percent of January admissions were because of breathing difficulties, 14 percent in February, 17 percent in March, and 52 percent in April. During the entire year, four hospitals admitted 233 people suffering from breathing problems—33 died. Some of the patients were moribund when admitted and died within a few hours. Some had traveled long distances to reach a hospital, frequently during a severe dust storm, and their resistance was lowered from a long ride in dust-laden air. Many hospitals distributed gauze masks to make breathing easier for the patients.[8]

In the spring of 1935 a journalist reported from Great Bend, Kansas, that "Lady Godiva could ride through the streets without even the horse seeing her." Schools and businesses closed during the worst dusters, and chickens went to roost in the early afternoon. A group of Dalhart, Texas, citizens, hoping to find a solution to the twin plagues of drought and dust, met at the county courthouse in late April to hear explosives expert Tex Thornton discuss the possibility of detonating dynamite in the air to bring rain. This scheme aroused the crowd's interest when they heard testimony crediting aerial bombing with breaking the 1892 drought near Council Grove, Kansas, and were assured that the rains in France during the war were caused by the nearly continual artillery bombardments. Some of those present also claimed that dynamite blasting in the caliche pits near Dalhart had brought an abundance of rain to the local area in 1934.[9]

Thornton was optimistic that explosives could bring rain if they were detonated on a day with low cloud cover and moderate temperatures. Farmers and ranchers guaranteed Thornton $300 for the purchase of TNT and solidified nitro-glycerin jelly which he proposed to set off at twenty-minute intervals. Equipped with time fuses, the explosives would be sent to the cloud ceiling in balloons with connecting strings to control their height. Thornton scheduled the blasting for May 1 near Rita Blanche Lake, four miles southwest of Dalhart. City businessmen scheduled a street dance for that evening to celebrate the results—in the rain if possible. At the appointed hour, several thousand farmers, ranchers, photographers, newsreel cameramen, and reporters gathered to watch the aerial bombing, but a sudden dust storm drove the onlookers to the protection of their cars. Thornton, though visibly upset, was intent on completing his rainmaking venture. Since the wind was too strong to send explosives aloft in the balloons, he buried the sixty charges in the sand and set the fuses. The resulting blasts threw additional dirt into the air, where it mingled with the blowing dust, causing even more discomfort to the audience. No rain fell, and Thornton post-poned plans for additional detonations.[10]

Not everyone, though, believed conditions in the southern Plains could be changed, and, thinking that they could not be worse off anywhere else, many chose to leave the Dust Bowl. In mid-April 1935 the Federal Emergency Relief Administra-tion reported that one hundred "normally self-sustaining" families had left Texas County, Oklahoma, in a month's time. In Cimarron County, Oklahoma, the dust had driven out all but three of the forty families that lived in the six townships south of Boise City. One Oklahoma woman loaded her pos-sessions, including the family goat, on top of a truck and with her three children headed for an uncertain future in Colo-rado. A Hardesty man packed up his family and moved to California saying, "We feared for the family's health if we stayed here. We couldn't make it go here anyway."[11]

Although John Steinbeck's Joad family will epitomize for all time the plight of those who were dusted out of the south-ern Plains, the vast majority of the people remained in the

Dust Bowl. What made them stay? Many remained because they had an overwhelming faith in the future. Three words— "if it rains"—dictated much of their daily life. If it rained, the wheat and pastures would grow. If it rained, the dust would settle. If it rained, the region would prosper again. Stubbornness made others stay. In 1925, one farmer living northwest of Guymon, Oklahoma, had $35,000 in the bank after farming for thirty years; ten years later and nearly without funds, he said, "I could have left here wealthy and I'll be damned if I'm going to walk out of here broke now."[12]

A farmer near Boise City, Oklahoma, had a different reason for staying. He noted, "If I leave, I can't get wheat and corn payments on relief and that's all that keeps me alive." Certainly, government aid in the form of subsidies, loans, and grants allowed many farmers to remain in the Dust Bowl. The support of a host of agencies such as the Federal Emergency Relief Administration, Agricultural Adjustment Administration, and Farm Credit Administration enabled farmers to buy tractor fuel, horse feed, seed, food, and clothing and thus to endure the menaces of drought, dust, and hard times until the rains returned. Some people no doubt remained because they were simply afraid to leave or had no place to go. Others stayed because they had raised their families in the southern Plains, and their homes held too much sentimental value to leave behind.[13]

Those who stayed fought the dust storms religiously and psychologically as well as physically. When a duster swept beyond the southern Plains in mid-March 1935 a Topeka, Kansas, woman called a city newspaper and warned that it was the "wrath for the second coming of Christ." Others quoted Deuteronomy after a black blizzard passed: "The Lord shall make the rain of thy land powder and dust; from heaven it shall come down upon thee, until thou be destroyed." For them, the storms were a form of visitation from the Lord. Gene Howe, editor of the *Amarillo Globe*, agreed: "No doubt there has been sufficient disobedience in this Western country to justify the Lord in almost anything he might do to us."

But he could not believe God would single out Dust Bowlers for vengeance since they were "so much better morally and religiously than the folks who live in the big cities in the eastern part of the United States that there is no comparison." For Howe, the big corruption was in the urban areas like New York, Chicago, and Philadelphia, and the people who lived there were more deserving of divine wrath than the citizens of the southern Plains.[14]

A number of fundamentalist religious sects increased the distribution of their church materials and held street corner meetings to warn people to be ready because "the storms were the fulfillment of the signs of the times and heralded the approach of the end of the world." One hardened Dust Bowl resident noted that only the "weaker type of people" listened to such propaganda. In Kansas, a Hodgeman County prayer group representing all faiths appealed to Dust Bowlers for "reconsecration to God and right living and to prayer for drought relief," and more than two hundred persons attended the first meeting of the group. The Hodgeman prayer group sent a petition to President Roosevelt and Governor Landon proposing thirty days of "earnest prayer" so that "God may come to our rescue speedily." In Colorado a Dust Bowl newspaper reported: "Prayer will be held at the Prairie Center Schoolhouse next Sunday, May 5[1935], at 2:00 o'clock on behalf of rain. We do not hesitate to ask God for other necessities so why not this?" A year later in March 1936, Elkhart, Kansas, residents organized a mass prayer meeting to be held on main street. All merchants were requested to close their businesses for it, and all of the town's pastors were asked to be present. The local editor wrote: "Regardless of church affiliation we are all worshipping the same God and let us unite in one earnest appeal to Him for relief." Other Dust Bowl localities held similar prayer services for rain.[15]

Certainly the clergy felt the economic impact of the drought and dust storms. Many people in the drought-stricken area stopped going to church because they could no longer afford adequate clothing. With the southern Plains economic and agricultural base nearly destroyed, there was little money available to support the churches. As a result, some pastors earned as little as $250 annually. Because of this distressing condition, the *Missionary Review of the World* published an article entitled "Churches in the Dust Bowl." In it, the author urged farmers to practice soil conservation and thereby to increase agricultural productivity and the region's economic well being. Specifically, he wrote: "The stewardship of the soil is a religious responsibility. . . . You cannot maintain good churches where the soil has been depleted."[16]

Humor was an incalculably important psychological release that helped people cope. Indeed, people reported the drought as so bad that when one man was hit on the head with a rain drop, he was so overcome that two buckets of sand had to be thrown in his face to revive him. Housewives supposedly scoured pans clean by holding them up to a keyhole for sandblasting, and sportsmen allegedly shot ground squirrels overhead as the animals tunneled upward through the dust for air. Some farmers claimed that they planted their crops by throwing seed into the air as their fields blew past and that birds flew backwards to keep the sand out of their eyes. One local report indicated that when a traveler stopped at a Dust Bowl restaurant and ordered a boiled egg, it was full of dust when cracked. Five more eggs were opened with the same result. At this point the traveler gave up trying to eat an egg, but took a dozen eggs home and put them under a sitting hen. To his utter surprise, the eggs hatched three weeks later, producing ten mudhens and two sand-hill cranes.[17]

Dust Bowlers supposedly learned to test wind velocity by tying a logging chain to a tree. If the wind blew the chain straight, the day was calm. If the chain popped like a whip, a heavy breeze was blowing. But, if the chain snapped to pieces and the tree was pulled up by the roots, it was a blizzard or a twister. On occasion, some residents claimed that the wind blew away so much soil that post holes were left standing above ground, and farmers loaded them into wagons and stored them for future use. Natives claimed they could accurately predict the approach of a duster or "Oklahoma rain" when the rattlesnakes started sneezing. One man reportedly put some bullfrogs in a water tank to multiply, but they drowned because they had never been in water before or learned to swim. Considering the extreme drought, one farmer mused, "I hope it'll rain before the kids grow up. They ain't never seen none." Kansans said they knew how to take a dust storm: "They take it on the chin, in the eyes, ears, nose, and mouth, down the neck, and in the soup." A Kansas City engraving firm advertised small bottles of water for sale to Dust Bowlers and noted, "It's been a long time since you have seen any of this stuff—real genuine rain water. This specimen has been especially imported by us and bottled in our own plant."[18]

One Texan demanded that people recognize that sand-storms were native products of the state—dust storms were not. "We are rather proud of our sandstorms," he wrote, "because they tell us that West Texas is on the move. These dust storms are intrusive aliens from Kansas and Nebraska." Furthermore, "a sandstorm in which flying particles do not achieve the size of a small hen egg is not a genuine Texas sandstorm but an imported variety. It is an intruder from the effete Middle West, and like other effete products a whole lot nastier than the big, roaring, open variety properly associated with the name of Texas."[19]

John L. McCarty, editor of the *Dalhart Texan,* reflected this humorous and obstinate spirit when he urged the Dust Bowl's inhabitants to "Grab a root and growl." In Dalhart, he organized the Last Man's Club—admission obtainable to each man who pledged that he would be the last one to leave the Dust Bowl. McCarty got the idea for the Last Man's Club after reading a news item about a Civil War unit that had organized a last man's club. McCarty applied the idea to the Dust Bowl and the club was organized on April 15, 1935, with McCarty as President. About one hundred joined; more wanted to do so but were afraid they might have to leave.[20]

McCarty tried to instill a sense of pride in the region on the part of his readers. Driving home from Sunray, Texas, one afternoon he observed an approaching black blizzard. The sight of it left a lasting impression on his mind. The storm occupied about a third of the sky. It was purple, blue and dark green as the sunlight struck it, and the sky around it was dark blue. For McCarty, "It was one of the most dramatic things you can imagine." When McCarty arrived home, he sat down at his typewriter and composed an article entitled "A Tribute to Our Sandstorms." He published it in his paper and other newspapers around the country picked it up. The article not only described the awesome beauty of a black blizzard, but it also helped to dramatize the plight of the Dust Bowl farmers.[21]

McCarty wrote: "We've got the greatest country in the world if we can just get a few kinks straightened out. . . . Let's keep boosting our country." With a friend McCarty jokingly planned to build a huge hotel in the midst of the sand dunes north of Dalhart. Their plan was to lure tourists to watch the "noble grandeur and imposing beauty of a Panhandle sandstorm." Of course, they planned to charge "fancy prices" for the privilege of the experience.[22]

The editor of the *Muleshoe* (Texas) *Journal* had this same spirit, and believed that a Panhandle sandstorm should be classified among the seven wonders of the world. Nothing could exceed its "terrific beauty," "majesty of manner," or "decisiveness of action." For this editor:

The Panhandle sandstorm will move more dirt in an hour than a hundred realtors can in several days. It will blow down more buildings than a regiment of carpenters can erect in a month. It will cut off more cotton in a day than Henry Wallace can get plowed up in a year. It will turn a chicken's feathers wrong side out quicker than a housewife can pick it for dinner. If it were not for the fact that the wind blows about as much in one direction as it does another, farmers would frequently have to travel many miles to cultivate the land that was really theirs,

and instances are on record where men have taken their teams
across the fence into the neighbor's field so as to plow up the
land they had actually bought.[23]

With regional pride very much on his mind, McCarty wrote
a blistering reply to Walter Davenport's article entitled "Land
Where Our Children Die" published in *Collier's*. McCarty
titled his response "Thou Shalt Not Bear False Witness," and
charged that the *Collier's* story was a "vicious libel." The
intent of it was, he maintained, to prove that the people of the
Dust Bowl were insane, parents of "unnatural children,"
given to homicidal fury (especially during dust storms), racke-
teers, political beggars, thieves, hysterical cowards, and
incompetent farmers—all for the purpose of "taking a crack"
at federal agencies which were sincerely trying to solve the
wind erosion problem.[24]

Contending to speak only for the "fine Americans" living
in the Dust Bowl, McCarty resented *Collier's* "swaggering air
of finality." He admitted that the region had its problems, but
he accused the magazine of "rotten reporting" and of "lies and
half-truths." All in all, he said the article was "perhaps the
dirtiest and most damaging journalism ever used against this
area by an American newspaper or magazine." McCarty suspected the article was written as propaganda for the Republican party. Another Dust Bowl editor responded differently,
noting that "John is mad because the nation found out that
things out here are bad, and because Walter [Davenport] says
they are worse than that. And so goes life and human nature."
McCarty wrote privately, though, that the effect of the story
in *Collier's* was beneficial because it publicized the needs of
the region.[25]

A colleague of McCarty's reflected the same regional pride
when he wrote, "We'll stay here, not 'till Hell freezes over'
but until we have convinced the world that we are proud of
our country, that it has given us all that we have, or hope to
have, that it is a country, under normal conditions, which far
surpasses any other on God's green earth."[26]

In 1936 photographers from the Farm Security Administration including Arthur Rothstein and Dorothea Lange, toured the Dust Bowl and recorded what they saw. The Resettlement Administration also sent a film crew to portray the causes and consequences of the Dust Bowl and to suggest what could be done about it. The product was a film entitled *The Plow that Broke the Plains.* Rexford Tugwell's purpose in producing the film was to convince Congress to grant funds for the agency's work in the southern Plains. It too drew the wrath of McCarty and many others like him. This film, directed by Pare Lorentz with music by Virgil Thomson, was hailed by some as a "work of art" and "thirty minutes of unforgettable pictures." Franklin D. Roosevelt endorsed the film as a "modern tool of government." Tugwell called it "America's first documentary film." It was previewed before congressmen, diplomats, and New Dealers at the Mayflower Hotel in Washington, D.C., on May 10, 1936. The film began with a rolling sea of grass extending as far as the eye could see. Soft music accompanied a commentator who explained how 40 million acres of Great Plains grassland looked before the coming of the pioneers. As the cattlemen and homesteaders arrived, the music grew louder. Gang plows began to turn the sod, wheat began to grow, and the first faint puffs of dust lifted into the air. Next, a newspaper headline announced the first World War and soaring wheat prices. On the horizon a row of tractors appeared, rapidly breaking more sod for wheat. The music was by then deafening. The camera flashed to a jazz band and to an exploding stock ticker. The screen went black. Within a few seconds, the scene had changed—a steer's skull appeared in the dirt, and the camera captured a house nearly inundated by dust. "Forty million acres of plains totally ruined by the plow," the commentator says, "200 million acres badly damaged. What is America going to do about it?" The lights flashed on; the film was over. Even Dust Bowlers had to admit that it carried a "terrific punch."[27]

T. E. Johnson writing for the *Amarillo Globe* noted two main faults with the film. First, it portrayed only the worst side of the Dust Bowl. Second, the film minimized the role of drought. As a result, the blame for the Dust Bowl was placed on the rancher who overgrazed the Plains, and upon the farmer who broke too much sod for wheat. It also left the impression among some people that sand dunes covered the southern Plains rather than existing only in isolated areas. Although Johnson thought the picture "impressive and technically correct," it was "by innuendo, misrepresentations and plain untruths a vicious and damaging document, produced by taxpayers money." More important, perhaps, was that "its showing will make Easterners call us boobs for living in the West and Westerners hang their heads in shame for letting Easterners run the country."[28]

Eugene Worley, a citizen of Shamrock, Texas, and a delegate to the Democratic national convention, demanded that the party prevent the Resettlement Administration from showing this film because it was a "libel on the great Texas Panhandle" and on the "most hospitable and courageous people." If Rexford Tugwell did not withdraw it from circulation, Worley threatened, "I'm liable to punch him in the nose." Two thousand theaters quickly booked the film, and it was distributed by independent theater chains over the opposition of the major Hollywood companies. The major distributors refused to handle the picture; they argued that it was too short for a long feature and too long for a short one. The real reason for not using it was that they resented the federal government's intervention and competition in the entertainment field. Actually government films such as *The Plow that Broke the Plains* were not new. The Department of Agriculture and many New Deal agencies often used this technique to convey information. None, however, had the shock effect or as much popular appeal as this film did.[29]

Dust Bowlers also criticized the picture because it was shown in New York City in the same theater where two Soviet films—*May Day in Moscow* and *News Spots from U.S.S.R.* were released by the National Russian Picture Exhibiting Company. One Amarillo businessman clearly saw the relationship of the film and New Deal to socialism and claimed that the showing was made in a theater operated by Amkinos, the Soviet subsidized film agency. Evidently, the film made more of an impression on Dust Bowlers and New Dealers than it did on New Yorkers. When the ticket girl was asked what the film was about she replied that it was "something about farming out west." Nevertheless, the film served the purpose and enabled Tugwell to get his rehabilitation program through Congress. But for some Dust Bowl residents it was "like the false teachings of Hitler [and] will not be erased from people's minds for a long time."[30]

Alexander Hogue also brought national attention to the Dust Bowl through his oil paintings. Hogue gained international recognition for his paintings such as "Drought Survivors," and "The Grim Reaper." *Life* magazine reproduced some of his Dust Bowl work in 1937. John L. McCarty complained that Hogue's scenes were not typical and portrayed the Texas Panhandle as being one vast desert. To this, Hogue replied, "That's what it will be if they don't let the government do its work." The West Texas Chamber of Commerce refused to admit that these paintings were works of art, criticized Hogue for showing the region in a negative fashion, and branded him "disloyal." Hogue responded by saying, "That's just what I expected from a chamber of commerce," and suggested the Dust Bowl "would be a lot better off if the chambers of commerce would quit interfering with what the government [was] trying to do." The West Texas Chamber of Commerce did not recant, and the Texas House of Representatives, prompted by delegates from the Panhandle, censured Hogue's pictures.[31]

At the same time that Hogue's paintings appeared in *Life,* the *March of Time* film company released a motion picture short featuring the Dust Bowl. The *March of Time* camera crews traveled six thousand miles on the roads near Dalhart and shot over twenty thousand feet of film in an attempt to reveal the conditions in the Texas Panhandle. The film was screened in Amarillo after which the Amarillo Junior

Chamber of Commerce wired a protest to the film company claiming that the documentary did not accurately portray the true conditions in the region. Eugene Worley again persuaded the legislature to vote a resolution of reprimand against *March of Time*. Worley argued, "I think they made a trip to the Sahara Desert to get some of those pictures." The scenes were not typical of the Dust Bowl, and "didn't show the good places." Ed Bishop, editor of the *Dalhart Texan*, believed that many years would pass before the adverse publicity of the film could be overcome. In an emotional response to the film he wrote:

> We need men in our government agencies who believe in this country. We need men in business houses, on the farms, in the banks, and in newspapers who will say "Take your movie cameras, your magazines, radios and newspapers and go to hell with them. This is our country, it has given us all that we have and all that we expect to have and to it we are grateful. We believe in it, live on it and expect to die here, happier, more content, more God fearing, better educated, and healthier, than any other section of a great country. We want none of your unfair, biased and exaggerated publicity. We gave you an opportunity to tell the true story and you answered with lies. All we ask of you now is to be let alone."[32]

Most Dust Bowl newspapers did not religiously record each passing dust storm. Most editors thought they were no longer news or that it was bad business to report them since it discouraged tourism and investment in the region. One local editor wrote that when reading about the plight of the Dust Bowl in the New York and Los Angeles newspapers "it appears that most of us [are] dead or dying." Another southern Plains editor correctly wrote: "Weather is like crime. The sensation is the evil, the malignant. When the bandit is calmly lying in jail he is no longer news. When the wind's quiet and the soil is being tilled normally, the weather is no longer news and likewise not worth wasting much ink on. The abnormal makes news."[33]

Some Dust Bowlers turned their stubborn pride into a blind faith in the region's renewal. One Baca County, Colorado, resident wrote: "With real pioneer courage we are doing our best to make the most of a bad condition. We are able to better teach our children lessons of gratitude and self-reliance because of the unfortunate experiences they are going thru. Boys and girls who face the dust-sands on their way to and from school for weeks will be stronger when facing other problems of life. . . . We feel toward our country as parents toward a wayward child. We love it in spite of the wind and drought. Only the faint hearted will move away. The pioneers will stay, plant trees and trust in God." Another wrote: "The first few years I lived in the Dust Bowl I felt I could not endure them. I would cry, laugh, and then get angry because I had not more sense than to stay in such a place. When I knew there were other places where I would never see a dust storm much less hear of one. But somehow I would go through them thinking of the beautiful day after the storm. Something like we look forward to the resurrection."

Others were even more resigned to living in the Dust Bowl. One farmer philosophically noted: "We could have pulled up and left, but we like the people and the country in general and of course we look for a better year next year and we feel sure other localities have their drawbacks and maybe greater than ours." Another Dust Bowl resident stoically professed: "When I see a dust storm coming, I feel that it is God's will—so I try to like it."[34]

In retrospect, physical and psychological stamina, humor, and regional pride enabled Dust Bowlers to endure, adapt, and survive in the harsh environment of the southern Great Plains. Although conditions were bad for most of the 1930s, the storms for the most part came in the spring. The remainder of the year was remarkably free from dust—blue skies and occasional periods of light rain prevailed. A western Kansas school teacher reflected her fascination with life in the Dust Bowl when she wrote, "Your education is not complete until you have lived with these people for a while. Only then

will you know what an optimist is. Yet, I am charmed with
this country, which today is beautiful with its quiet wide-
spreading vistas, and tomorrow may bring anything in the cat-
alog of possibilities." For her the Dust Bowl was a "glorious
country" which at its best and worst moments had no equal.
Another resident wrote, "We have faith in the future. We are
here to stay."[35]

In May 1936, Caroline A. Henderson, a twenty-eight-year
resident of Alva, Oklahoma, wrote a series of letters to a
friend in Maryland which were published in *Atlantic* magazine. The elegance, grace, melancholy, and optimism
reflected in those letters clearly indicates that she was really
writing for posterity. She wrote: "We long for the garden and
little chickens, the trees and birds and wildflowers of the
years gone by. Perhaps if we do our part these good things
may return someday, for others if not ourselves." Whether
staying in the Dust Bowl reflected "courage and perseverance" or "recklessness and inertia," she could not say. But,
"to leave voluntarily—to break all these closely knit ties for
the sake of a possibly greater comfort elsewhere seems like
defaulting on our task. We may *have* to leave. We can't hold
out indefinitely without some return from the land, some
source of income, however small. But I think I can never go
willingly or without pain that as yet seems unendurable."
While Caroline Henderson and others like her stayed, they
worked hard to restore the land.[36]

CHAPTER FIVE

Soil

Conservation *

Prior to 1930, American farmers commonly regarded soil erosion as a "spasmodic phenomenon" largely confined to lands subject to washing. Many farmers believed moderate blowing was beneficial because it mixed the soil and thereby helped to maintain fertility. The average agriculturist seldom did more to contend with the problems of erosion than to haphazardly place brush in gullies. Urbanites understood the problems of erosion even less than farmers did. The fourth decade of the twentieth century quickly challenged such carelessness. The drought which began in the summer of 1931 lasted seven years and made plainsmen acutely aware that they had to change drastically their agricultural methods.[1]

Dust storms and drifting soil were by no means new experiences for Dust Bowl farmers. Neither was the technology they used to bring the soil under control unique. At the turn of the century, farmers in the Oklahoma Territory used the lister and the disk harrow to check soil blowing. The farmers in western Kansas also used the lister to control blowing dust

---

* This chapter was previously published, in different form, as "Agricultural Technology in the Dust Bowl, 1932—40," in *The Great Plains: Environment and Culture,* edited by Brian W. Blouet and Frederick C. Luebke (Lincoln and London: University of Nebraska Press, 1979). Copyright© 1979 by the University of Nebraska Press.

in 1913. Farmers also left stubble on the land when blowing was a problem, or, if it had already been turned under, they spread straw across the fields.[2]

During the great plow-up of the 1920s, though, Great Plains farmers largely ignored the lister and turned to the one-way disk plow because it plowed faster, handled heavy stubble without clogging, efficiently broke hard, sun-baked soil, and destroyed weeds. However, if used continuously on fallowed ground, the one-way disk plow pulverized the soil and subjected it to wind erosion. Repeated cultivation will retard soil blowing only as long as the surface is kept in a rough condition. Some Dust Bowl farmers were reluctant to use the proper tillage implements, particularly the lister, when their fields began to blow. These skeptics believed that any further tillage would stir the soil and intensify the problem. One Kansan wrote: "I do not take much stock in the effort being made to plow or list the ground to prevent blowing; there is too much territory to be covered effectually by such puny efforts, even if it were a sure thing that favorable results would follow. Even if the state and the government together had sufficient machinery and manpower to go over enough territory to accomplish results, by the time the organization is complete and they get started to work, the windy season will be over." He regarded an emergency listing program as a "mere gesture" to prevent dust storms, and one that would accomplish "very little if anything" in the end. Nor did some farmers believe the wind could blow all of the top soil away. One farmer living northwest of Guymon, Oklahoma, reflected the carelessness of many Dust Bowl agrarians when he said: "Let the wind blow the top soil away. We can plow up more. . . . You just can't seriously hurt this land out in the Panhandle."[3]

By the spring of 1934, wind erosion in the Dust Bowl was so serious that most farmers and ranchers were willing to adopt the appropriate measures to bring their soil under control, and the lister became one of the standard implements used for such purposes. Some farmers preferred a modified lister which incorporated a shovel attachment that built small earthen dams in the furrows. This "basin" or "damming" lister enabled twice as much moisture penetration as on land treated with the standard lister, but unfortunately it did not work well on dry soil. The Fort Hays experiment station in western Kansas demonstrated the basin lister during the spring and summer of 1938. Over three thousand farmers in thirty counties watched the demonstrations. The experiment station also distributed over four hundred blueprints which showed farmers how to convert their standard lister into the "damming" model. Some farmers obtained a similar effect by shearing off a portion of a one-way plow's disk so that it pocked the soil instead of making a clean cut across a field. Other farmers modified their one-way disk plows by removing all but four evenly-spaced disks; this adjustment did not furrow the soil as deeply and enabled the farmers to level their fields more easily after the blow season. Dust Bowl farmers also used a combined lister-seeder for planting row crops. Even though its plantings started slower and produced less, it was the cheapest planting method and greatly reduced the wind erosion hazard. If wheat was planted in listed furrows the ridges could be cultivated back into the furrows when the plants were tall enough to hold the soil. Although some of the wheat was killed during this process, the soil treated in this manner held more moisture and enabled greater yields at harvest time.[4]

In the spring of 1935, the Federal Emergency Relief Administration financed a massive soil listing program. Largely due to the efforts of Governor Alfred M. Landon, Kansas received $250,000. This money was distributed to farmers through the Kansas Emergency Relief Committee. Farmers received ten cents an acre for emergency listing of their blowing fields (later increased to twenty cents an acre in 1936), and forty cents an acre if they had to hire the work done. In Hamilton County, farmers were allotted one gallon of fuel and one-sixteenth gallon of oil for each acre worked with tractors, and ten pounds each of grain and hay for every acre worked with horses. The intent of the program was to put 15,000 tractors to work blank listing 4.5 million acres and strip listing an additional 2 million acres of Kansas land. County commissioners and agricultural agents were to supervise the listing program and determine which lands should be worked. Farmers and tenants were to work without pay and use the federal money only to buy tractor fuel and horse feed. In order to benefit local communities in thirty-nine western Kansas counties affected by this program, fuel and feed purchases had to be made from local dealers. Farmers were given 60 percent of the money necessary to list their fields in advance. When the work was satisfactorily completed, they received the remaining 40 percent.

Some farmers balked at participating in the listing program because they had to declare themselves paupers when applying for funds. Most farmers favored the plan and were eager to begin the conservation work. As a result of that work, almost all of the May rains in 1935 soaked into the soil and facilitated the growth of a considerable feed crop even though the drought continued. Vegetative cover also began to grow on the listed land and contributed to stabilization of the soil during the next two years. The Texas Panhandle also received $500,000 for emergency listing of 5,395,000 acres, and New Mexico received $400,000 for the same purpose. This federally financed program was designed to begin soil conservation practices until a national program could be formulated.[5]

The following year, Congress appropriated an additional $2 million for spring emergency listing. Although the severe wind erosion area had been reduced in Kansas to half that of the previous year, soil was still blowing in the Dust Bowl counties. As a result, forty-two Kansas counties received $500,000; Colorado received $530,000 for emergency listing in twenty-one counties; New Mexico got $250,000 for seventeen counties; Oklahoma received $220,000 for six counties; and, Texas got $500,000 for twenty-five counties. Out of the $2 million appropriation, $1,422,591.53 was spent in the spring of 1936. The balance was kept in the state treasuries for use the following spring. By the end of June 1937, $1,853,948.28 had been spent for listing 8,028,642 acres; an additional 1,903,025 acres were listed at farmers' expense. All states except Kansas reported wind erosion conditions "greatly improved" because of the program. With the depletion of the $2 million fund by the summer of 1937, the emergency listing program ended, and the federal government rejected subsequent requests for similar funds.[6]

The lister, however, was not the only useful implement for controlling soil blowing in the emergency listing program. Virtually any tillage implement that roughened the ground was effective. The duckfoot cultivator and the rotary rod weeder were particularly good on summer fallow. The duckfoot cultivator not only killed weeds but also ridged the soil and kept the trash on the surface. The shovels cut beneath the surface and did not pulverize the soil. The edges dulled quickly, but this problem could be overcome by treating them with a hardening material. The rotary rod weeder consisted of a square rod which ran approximately two inches beneath the ground surface and rotated backward. In so doing it cut weed roots, kept trash on the surface, and left the soil intact. It was especially effective for preparing the seed bed because it leveled the ground and permitted greater uniformity in planting depth.[7]

The spring-tooth harrow was useful in controlling wind erosion, but it gathered trash badly and was more effective when subsurface shovels replaced the harrow teeth. Some farmers used the subsurface packer equipped with a series of narrow, wedge-shaped disks that broke crusted soil and created deep crevices which caught windblown soil and retarded further soil movement. All lands subject to blowing needed periodic treatment with these implements. Listing, for example, held the soil during the blow months for about one week—then the work had to be repeated.[8]

Limited use of these implements could not markedly improve the wind erosion conditions in the Dust Bowl; they had to be used widely and properly. For example, if one farmer listed his fields while his neighbors did not, the listed fields would have little or no effect on soil blowing. In addition, the implements had to be used in such a fashion as to insure maximum benefit from every drop of rain that fell. This was particularly important when approximately three-fourths of the annual precipitation fell between April and September. The southern Plains farmers failed to conserve this moisture properly in 1934, and, as a result, they harvested Russian thistles for cattle feed; in 1935, when the thistles had dried up, they cut soap weed.[9]

The semiarid, dusty conditions prompted the Soil Erosion Service (SES) and the Department of Agriculture to stress the benefits of contour plowing and terracing. Hugh Hammond Bennett, director of the Soil Erosion and Moisture Conservation Investigations, recognized that technical supervision of such a program would be necessary, and Harold Ickes, Secretary of the Interior and head of the Public Works Administration, provided help on August 25, 1933, when he allotted $5 million of emergency funds to the Department of the Interior's Soil Erosion Service for a conservation program. The SES used this money to establish demonstration projects on private lands where cooperating farmers signed five-year contracts in which they agreed to follow the conservation practices which the SES inaugurated. The SES hoped these projects would stimulate all Dust Bowl farmers to begin similar conservation practices. The first wind erosion demonstration unit established was a 15,195-acre project east of Dalhart, Texas. Originally, the federal government allotted $35,000 for the project, but by October 1934 that amount had been doubled and the project had been expanded to 30,000 acres. Similar projects were soon established in Colorado, Kansas, Oklahoma, and New Mexico and in several other Texas Panhandle locations. In addition, the Civilian Conservation Corps (CCC) established fourteen demonstration projects in the Dust Bowl. Farmers within twenty-five miles of the CCC camps could sign cooperative agreements for soil conservation work. Each farmer was expected to pay for as much of that work as possible, to maintain terraces and other improvements, and to observe recommended soil-use plans for five years.[10]

Within the demonstration projects, the SES, county agents, and the CCC provided the necessary technical expertise and equipment to encourage farmers to participate in a conservation program. Contour plowing was one technique they advanced as basic to soil conservation. It enabled the soil to retain a higher percentage of moisture and permitted greater vegetative growth than was possible on land not worked on the contour. But the spring of 1936 saw little contour farming outside the demonstration projects. During that year, however, federal aid enabled Dust Bowl farmers to list 4,469,270 acres—2,469,534 acres of which were contoured. Tests on contoured lands following late May rains indicated that about one inch of additional moisture was conserved and subsoil moisture was more than a foot deeper than on lands not contoured. These results prompted soil scientists to predict the area's grain production would increase 4.5 million bushels and yield 500,000 more pounds of crop residue for a protective soil covering. A Texas farmer north of Amarillo also proved the benefits of contouring when he raised 160 acres of feed on contour plowed land. No one else in his neighborhood had contoured their land and no other feed grew. Similarly, in Randall County, Texas, farmers spent $18,000 to list 120,000 acres of wheat in the spring of 1936; they harvested 75,000 acres for a total wheat income of $300,000.[11]

Because the SES duplicated many of the activities of the Department of Agriculture and because of the pressure from the extension services of the land grant colleges, the SES was incorporated into the U. S. Department of Agriculture. The SES was then rechristened the Soil Conservation Service (SCS). That same year, the Soil Erosion Act committed the federal government to a massive conservation program; it did not, however, provide for the implementation of the plan. The SCS quickly found that although demonstration projects

did persuade farmers to adopt proper conservation techniques, the projects seldom had influence beyond fifty miles
of the boundaries. In short, if conservation was to restore the
Dust Bowl, technical expertise and information had to be dis-
seminated on a far broader scale. The SCS did not believe the
federal government had the constitutional authority to
impose land-use regulations, but it did hold that the states
could do so. Therefore, the SCS began to stress that an
effective soil conservation program required the states to initiate land-use regulations and encouraged such action by
drafting a model state law on May 13, 1936, entitled *A Standard Soil Conservation District Law.* This statute provided for
the creation of state conservation districts by local petition
and referendum. By September 1, 1937, Colorado, Kansas,
New Mexico, and Oklahoma had followed the federal government's lead by passing similar laws. After a district was
organized under the guidance of the state soil conservation
committee, the farmers within its boundaries joined in unified efforts to combat soil erosion and misuse of the land.
District supervisors carried out various conservation practices, extended financial aid to farmers, signed contracts with
them, bought lands for retirement, and formulated land-use
ordinances subject to farmer approval. By June 30, 1939, the
Dust Bowl states had created thirty-seven conservation districts covering 19,036,000 acres.[12]

Texas was the only Dust Bowl state that did not pass the
model soil conservation law. In May 1935 the Texas legislature authorized the creation of local wind erosion conservation districts. These districts were to be identical to the
boundaries of counties and could be established by the vote
of local taxpayers. Under the Texas statute, the governing
body of each district consisted of the county commissioners
and county judge. The wind erosion districts were to prevent
damage to lands and highways due to drifting soil, engage in
soil conservation practices, borrow money, and accept monetary gifts and advances from the federal government. The
governing body had the power to treat blowing land that was

damaging other property and to assess the cost against the negligent farmer. This assessment constituted a valid first lien and was payable by the owner in three years at 5 percent interest. The governing body was to cooperate with the director of the Texas experiment station, the county agent, the state department of agriculture, and the federal government to halt wind erosion. Conservation programs were paid for by assessments on the treated land, from automobile registration fees, special taxes which voters might authorize, selected state tax money, and federal monetary aid.[13]

As early as 1913, the Kansas state legislature provided a similar law which empowered county commissioners to hire soil conservation work done on a negligent farmer's land. This power was widely used in the Colby, Kansas, area during the great blowout of that year. At first, the law applied only to counties of less than ten thousand population. In 1935 it was extended to all counties in the state. At that time, the law authorized county commissioners to do whatever was necessary to stop soil drifting. The expense incurred would then be assessed against the lands worked. The law did not provide any details for administrative procedure or standards of action.[14]

Execution of this law was hindered because emergency tillage costs were usually not included in the regular state budgets and had to be financed by "no fund" warrants. Some counties refused to honor those warrants until sufficient tax revenues had been collected to meet them. Consequently, the uncertainty or delay in payment for soil conservation work was a major problem in preventing the law from being an effective force against wind erosion.[15]

In late May 1936, the Kansas Supreme Court held the 1935 law unconstitutional. The case resulted from a complaint made to the Morton County board of commissioners concerning certain blowing lands. The owner did not obey an order to take immediate action to stop the soil drifting, and the board had the work done and assessed the charges to his taxes. In order to test the constitutionality of the law, original proceedings in *quo warranto* were brought on the grounds that the action taken was in violation of the federal Bill of Rights and the Kansas Constitution.[16]

The state supreme court declared the law invalid because the state legislature attempted to deal with a state-wide prob-lem by delegating undue legislative authority to county officials. This action, the court reasoned, permitted a lack of uniform regulation across the state. As such, the law did not ensure that county commissioners would follow sound principles of administrative law. The legislature remedied that defect in 1937 when it established explicit standards and procedures for counties to follow when treating blowing lands as well as for paying for the work done. This law provided for an annual levy of no more than one mill on the land owners of each county. The tax, however, could not exceed one dollar per acre annually. These tax revenues would be used to finance a forced tillage program on blowing, private lands, and for the planting of perennial grasses, trees, and shrubs upon the order from the county commissioners.[17]

Not everyone in the Dust Bowl believed that the SCS, the demonstration projects, and the soil conservation districts would be a success. One Kansan supposedly remarked, "If God can't make it rain in Kansas, how can the New Dealers hope to succeed?" In New Mexico some farmers referred to the federal government's plan to spend $10 million to buy 1,282,522 acres in Kansas, New Mexico, Oklahoma, and Texas in order to return it to grass as an absurd boondoggle where $30 an acre would be spent to restore land worth only $5 an acre. The *Albuquerque Tribune* was more favorable to that plan and noted: "Whether the methods to be followed are correct or not, the nation must have no qualms about a mere $10,000,000 experiment, the outcome of which may save America's breadbasket from desert. This effort to save the heart of American agriculture is not merely a New Deal policy. It is little less than a battle for survival."[18]

Certainly, the SCS had confidence in its program—particularly on the need to build terraces. Few terraces existed in this region even though the agricultural experiment station at Spur, Texas, had demonstrated by the summer of 1932 that a two-inch rain could be converted into a seven-inch rain when terraces were combined with contour farming. In December 1932, less than 10 percent of the nearly 4 million crop acres in a twenty-one county area surrounding Amarillo, Texas, were terraced or contoured even though terracing increased yearly earnings $2.00 per acre and raised land value $8.26 per acre. In 1935, terracing costs averaged only $2.25 per acre and increased land value $10.54 per acre; terracing was worth even more than this when a farmer mortgaged his land.[19]

The Soil Conservation Service continually worked to speed and improve the contouring and terracing process. Experiments at the Spur and Goodwell stations increased soil moisture an average of 25 percent on terraced and contoured land. The SCS estimated that such moisture savings increased the chance of a successful wheat crop about 75 percent and increased the average yield 35 percent. In 1934 at Goodwell, terraced plots which received only 3.4 inches of rain throughout the entire growing season produced 185.5 percent more hay than did unterraced land. Three years later at the Dalhart station terraced fields yielded 723 pounds of sorghum per acre while fields only contoured produced 589 pounds per acre. Here the advantages of contouring and terracing were clearly proven. Where the SCS planted sorghum in straight rows regardless of land slope, the yield was only 461 pounds per acre. The increased yield on terraced lands over those from unterraced fields was 262 pounds per acre or 56 percent, and the increase on contoured fields was 128 pounds per acre or 27 percent. In Carey County, Texas, a cotton farmer planted fifty terraced acres which produced twenty-eight bales. On forty adjacent, unterraced acres he produced only ten bales. The unterraced land averaged 125 pounds lint per acre and the terraced land 280 pounds lint per acre—155 pounds per acre difference.[20]

In May 1936, Baca County in the southeastern part of Colorado received the first beneficial precipitation since

September 1935. The amount of rainfall varied from 1.05 to 2.33 inches and provided the first significant test for the demonstration work conducted under the auspices of the SCS. The Springfield Civilian Conservation Corps camp reported moisture penetrated 41.67 inches on terraced lands but only 11 inches on unterraced fields. On land where no soil conservation practices were in effect, the runoff reached 90 percent. With little moisture retained in the soil, conditions were soon favorable for blowing, and ten days after the rain a severe dust storm struck the area. The message was clear—terracing and contouring held moisture in the soil, stimulated plant growth, and decreased wind and water erosion. By the late summer of 1936, fourteen SCS demonstration projects had 110,980 acres contoured and 33,021 acres terraced.[21]

The Soil Conservation Service also emphasized strip cropping in order to reduce wind erosion. Strip cropping consisted of planting a close-growing, soil-holding crop such as wheat, alternately with contoured strips of densely growing feed crops such as sudan, cane, sorghum, or small grains. Grain sorghum was particularly effective for stabilizing a field, and sudan grass grew rapidly and provided a dense wind-resistant growth. Both plant varieties were drought-resistant. Dust Bowl farmers found that sorghum, when planted in strips, protected wheat and fallowed land, collected snow, checked soil drifting, reduced evaporation and transpiration between strips, and provided an adequate soil-holding stubble and root system. The width of the strips depended on the tendency of the soil to erode, the condition of the summer fallow, the type of crop being grown, the wind velocity, and the amount of rainfall.[22]

Generally, Dust Bowl farmers realized success when they planted half of their fields with strip crops and utilized sorghum strips along fence rows to stabilize large drifts. The sorghum strips enabled farmers to remove the fences and stir the drifts so that the wind could redistribute the soil. Although strip cropping alone could not end the wind erosion hazard in the Dust Bowl, when farmers combined it with rough tillage practices and contour farming and terracing, it significantly checked soil blowing.[23]

In order to halt dust storms completely, though, grazing lands had to be restored. Not since the arrival of the settlers had the grasslands been in such poor condition. To help correct that problem, the Department of Agriculture, the Forest Service, and the Soil Conservation Service stressed grazing management, the reseeding of grasslands, and moisture preservation. The SCS advised farmers to rotate and rest pastures, to return at least 25 percent of the vegetative growth to the soil, to graze short grasses no less than two inches from the ground, and to restrict livestock to approximately twenty to thirty head per section of grazing land. Pasture rotation would permit the grasses to recover and produce seed. Although approximately 65 million acres in the Dust Bowl remained in grass through the 1930s, the carrying capacity of the range was far below normal. Overgrazing and drought had decreased the height and density of the native gramma and buffalo grasses to the extent that the soil had been completely denuded in some areas. E. A. Sherman, associate forester, warned that unless farmers practiced more conservative grazing management, soil blowing from the grasslands would help to create a desert in the southern Plains. By December 1934, nearly the entire native grass cover near Las Animas, Colorado, was smothered from the drifting soil, and by spring 1935, the pasture lands in western Kansas were 35 percent below normal in growth. Similar lands were drifting badly in Texas and Oklahoma.[24]

Dust Bowl farmers also practiced contour furrowing and contour ridging of grazing lands to derive maximum benefit from precipitation. The processes were not yearly operations and could be done efficiently with the implements readily available to Dust Bowl farmers—the lister, moldboard plow, or chisel. The contour furrow and the contour ridge differed slightly; the plow scattered the furrow slice as widely as possible while the ridge was left intact or even built up from overlapping two or more slices. The scattered furrow slice slowed runoff by allowing water to flow uniformly across the entire field, thus eliminating the possibility of washing and gullying. Since the furrow was beneath the surface, it was not subject to breaking and so had the added advantage of requiring no maintenance. Furthermore, grass runners extended down the furrow more rapidly than over the contour ridge, and seeds collected in the furrow enabling rapid vegetation.[25]

In contrast, the contour ridge served to impound water on the grass behind it; if the ridge was above the furrow, it spread the runoff over a wide area. The furrow checked water loss if the ridge broke. However, the contour ridge required a large amount of maintenance and was slow to revegetate. Contoured pastures that were not overgrazed could hold virtually all of the precipitation that fell and approximately 50 to 100 percent better grass growth resulted, depending on the sod condition prior to contouring and the adjacent blow areas.[26]

About 5 million acres, or 18 percent of the total acreage under cultivation in the Dust Bowl, were submarginal lands—lands that, given the current price of wheat, did not provide sufficient returns to meet a farmer's expenses from working it. Submarginal lands were often the first to suffer severe wind erosion. The SCS recommended that all of this land be revegetated with native grasses either naturally or artificially. The length of time needed for natural revegetation varied greatly depending on the amount of time the field had been broken, the proximity of seed grasslands, and the amount of grazing and rainfall received. Grassland experts estimated that twenty-five to forty years would be necessary for the more desirable grasses fully to recover abandoned fields. Farmers and ranchers also observed that fields which had been broken for only two or three years revegetated much more quickly than lands that had been farmed for a considerable period of time.[27]

The first task in achieving either natural or artificial revegetation was to stabilize the blowing land. If natural revegetation was desired, the goal was to obtain a weed growth as soon as possible which would hold the soil until more desirable vegetation replaced it. On the other hand, artificial revegetation required that weed growth be minimized while seed grasses became established. To minimize weed growth, the SCS advised farmers to plant a cover crop such as sudan grass, sorghum, or broom corn and to allow it to remain on the land during the winter. These crops protected the soil from blowing, acted as a moisture-conserving mulch, delayed weed growth, and protected the young grass plants after seeding in the spring.[28]

The SCS suggested seeding a mixture of blue gramma, buffalo, side-oats gramma, galleta, and sand drop seed grasses on heavy soils and a mixture of blue gramma, side-oats gramma, sand-blue stem, and drop seed grasses on sandy soils. Ten pounds of blue gramma and five pounds of side-oats gramma mixed with one pound each of the other grasses per acre produced the best results on heavy soils. Six pounds of sand-blue stem and one pound each of the other grasses planted on each acre was more successful on the sandy soils. When a Dust Bowl farmer chose between natural or artificial revegetation, he considered such factors as land value, the cost of seeding operations, the estimated length of time to reestablish grass by the various methods, the effect of neighboring blow lands, and his financial condition.[29]

Regrassing denuded pasture and submarginal lands was of crucial importance for stopping soil movement. Experiments indicated that where little blue stem, western wheat grass, and other mid-grasses held the soil, wind velocity on the soil surface was only 0.5 to 1 mile per hour, even though its velocity was 20 to 30 miles per hour above the ground surface. Nevertheless, during most of the decade, efforts to revegetate grasses artificially on submarginal lands met with limited success because rainfall was so far below normal that the grass seed usually failed to germinate and blew out of the soil. Furthermore, many farmers preferred to chance planting feed or wheat because these crops returned higher profits than did grazing lands. As late as the spring of 1938, Russian thistles contributed more to the stabilization of submarginal lands than did planned revegetation.[30]

Sand dunes posed a problem for Dust Bowl farmers in many areas, but particularly in Curry County, New Mexico, and Seward County, Kansas. The dunes were not only detrimental to farmers who owned the sandy lands but also to neighboring farmers, since the Federal Land Bank would not grant loans on property which was in danger of being covered by sand. Little was done about the sand dunes until early in 1936 when the SCS began experimenting near Dalhart and found that the most effective way to stabilize the dunes was to reestablish plant cover. The SCS recommended breaking down the steep slope on the leeward side of the dune with a tractor and disk or with a horse-drawn drag pole. Such a reduction in the height of the slope prevented the wind from forming eddies and allowed the particles of sand to blow beyond the dune. The dunes and the land between them could then be listed to help check soil movement and permit weeds, especially Russian thistle, to gain a hold. A cover crop of sudan grass, broom corn, kafir corn, or hegari gave extra protection. If such a procedure were followed, almost total stabilization and elimination of sand dunes could occur in a year—even under drought conditions.[31]

In July 1936, while the sand dunes were being broken down, President Roosevelt approved the transfer of $1.5 million from the Works Progress Administration (WPA) funds to the SCS for application to work projects involving soil conservation in the drought-stricken states. Of that money, eighteen Oklahoma counties received $125,000, while Colorado received $60,000. This money was spent under WPA regulations for the employment of certified drought-relief farmers. These men were to be employed in the construction of farm ponds and terraces. In Colorado, the work began in mid-November 1936; a month later 679 men were at work in thirteen counties. Work crews were also transferred from other WPA projects as fast as work sites could be found and equipment located.[32]

Work sites were determined, in part, by the proximity of the labor supply. Stock-watering dams were built within fifteen miles of WPA workers' residences. Farmers supplied equipment when possible; otherwise it was borrowed from the SCS conservation sites in the state or from the Forest Service. Since almost every worker was a farmer or stockman who had suffered considerable drought losses, they were eager to make the program succeed. Because that money had to be spent by the end of the year, time was a critical factor that threatened to restrict the program. Nevertheless, by the end of December 1936, 137 farm ponds had been constructed in Colorado. When the money was exhausted, all drought-relief workers were transferred to the Resettlement Administration or dropped completely if they were no longer needed. Resettlement Administration funds were not available to the SCS for support of its drought relief program, and the Colorado dam program ended before it could reach fruition.[33]

Dam building in the Dust Bowl was not the sort of useless "made-work" often associated with the Works Progress Administration. Even though drought conditions prevailed, occasional torrential rains caused heavy flooding, great property damage, and loss of life. In early June 1935, the worst flooding of the Republican River occurred immediately north of the Dust Bowl. The river crested twelve to seventeen feet above flood stage in portions of Kansas, Nebraska, and Colorado. Had farmers followed adequate water conservation practices, the runoff and subsequent flooding might have been less severe, though still damaging.[34]

Some Dust Bowl residents believed a federal bureau of aridity was needed to coordinate the conservation and drought relief work in the southern Plains. In late April of 1937, a group of Texas, Oklahoma, and Kansas farmers and businessmen met in Guymon and suggested that the limits of the entire Dust Bowl be specifically defined and that martial law be declared within them. They assumed that all cultivated land could then be forcibly subjected to the appropriate

conservation methods. Kansas' Governor Huxman refused to
request President Roosevelt to declare martial law. Huxman
did not believe that such action was a practical or possible
solution to the drought problem. In 1937 another Dust Bowl
resident suggested the creation of a Dust Bowl Authority saying, "There have been enough boy scouts and theorists, who
never did a lick of work in their lives, running around over
the country at government expense to have made a respecta-
ble standing army, and now after four years of monkeying
around, our country is [worse] than it was before they came."
A Dust Bowl Authority, he assumed, would give the needed
coordination to the various conservation programs as well as
help to develop a permanent conservation program.[35]

Unusual conditions pushed people to unusual solutions
which sought centralized authority. Normally, such authority
was absolutely contrary to the principles and beliefs of most
Dust Bowl farmers. Perhaps because of an increasing number
of suggestions to create a Dust Bowl Authority, the Southwest Agricultural Association, meeting in Boise City, Oklahoma in mid-May 1937, urged Congress to create such an
agency. Oklahoma Congressman Phil Ferguson introduced a
bill to create a Dust Bowl Authority. This agency was to be
located in the Dust Bowl; it was to be administered by an
assistant secretary of agriculture; and it was to be staffed by
"actual and bona fide" residents of the southern Plains. Secretary Wallace disapproved of a Dust Bowl Authority, and saw
no need to create a little TVA for the southern Great Plains.
Therefore, nothing came of this plan. However, in June 1937
the United States Department of Agriculture (USDA) con-
solidated all branches working on the wind erosion problem
in order to facilitate the development of a unified conservation program.[36]

By mid-December 1937, Dust Bowl farmers had reduced
the amount of seriously eroded land 65 to 70 percent from
the previous year. During the summer of 1938 good rains fell
over most of the Dust Bowl and allowed farmers to plant soil-
holding crops so that by spring 1939 the Dust Bowl had

shrunk to the smallest area since 1932. Although dust storms continued in 1939, the wind during the blow months was of less velocity and of shorter duration than at any time since 1934; the lands primarily subject to blowing at this time were those with sandy soils and denuded pastures. Nevertheless, more than 20 million acres of cultivated land needed terracing, and 3 million acres of crop land as well as 34 million acres of grazing land were not yet contoured. Ample rainfall occurred, though, in the summer and autumn of 1940, and for the remainder of the decade the Dust Bowl (and most of the Great Plains) received above average precipitation. With the return of the rains, the wind erosion problem temporarily ended.[37]

But the rains were not solely responsible for the restoration of agriculture in the Dust Bowl. The work of the Soil Conservation Service along with federal dollars were instrumental in helping farmers to bring their lands under control. By 1939 most farmers, following the technical advice and practices of the SCS, were using the proper tillage implements, were allowing crop residues to remain on the land during the blow season, and were contour plowing, terracing, strip cropping, planting drought-resistant crops, and following better grazing management practices. The farmers embraced the SCS programs because they were geared to practicality and low cost. For example, strip cropping was a simple farming procedure and did not require additional expense. The average cost for cultivating contoured land was twenty cents to fifty cents per acre, and the cost of building terraces was about $2.75 per acre. Most costs were minimal, since the SCS and other agencies provided thousands of dollars to help farmers initiate soil conservation programs; also, because the farmers had the necessary implements, they could do most of the work themselves. The SCS estimated costs for a farmer participating in its five-year program at $1.00 to $3.00 per acre. Furthermore, choice land in the SCS demonstration project near Hereford, Texas, which had sold for $8.00 an acre in 1935 increased to as much as four times that value by 1938. The tremendous increase in valuation came because those lands were once again stabilized and productive. Tests at the Goodwell, Oklahoma, agricultural experiment station indicated that land that had been terraced and contoured returned a net annual increase of $1.75 per acre.[38]

By 1940, 89 percent of the farmers under SCS contracts credited their conservation practices with increasing land value, 80 percent credited it for increasing their net farm income, and 95 percent intended to continue the programs after the expiration of their contracts. Farmer interest in the formation of conservation districts was high. So much progress had been made on the SCS demonstration projects that the staff which had previously been necessary to maintain operations was reduced to one or two technicians in fifteen of the seventeen projects.[39]

A combination of factors created the Dust Bowl in the southern Great Plains—overexpansion, cultivation of submarginal lands, failure to change crops when conditions required, lack of soil conservation practices, drought, and the relentless wind. The resulting dust storms of the 1930s forced farmers to utilize all of the technical expertise they could command to bring the wind erosion problem under control. When drought and dust storms returned to the southern Plains in the 1950s, the technology and conservation practices which Dust Bowl farmers had been using for the previous two decades prevented the region from reverting to the severe conditions of the 1930s. The consistent and proper use of technology and the attention to adequate soil protection were significant factors which enabled the Dust Bowl farmers to adapt, survive, and prosper in the harsh environment of the southern Great Plains.[40]

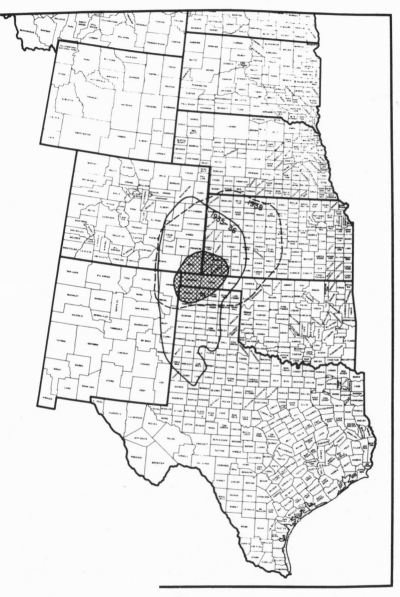

Solid and broken lines indicate the general boundaries of the Dust Bowl from 1935 to 1938. The checkered region received the most severe wind erosion. (Smithsonian Institution.)

A dust storm over Baca County, Colorado, brought total darkness for half an hour. (Western History Collections, University of Oklahoma Library.)

Heavy black dust clouds rolled over the Texas Panhandle in March 1936. (Library of Congress.)

Springfield, Colorado, was hit by a black blizzard on April 14, 1935. (Library of Congress.)

Boiling dust clouds helped christen the southern Plains the Dust Bowl. (Kansas State Historical Society.)

Dust storms often forced drivers to turn on automobile lights at midday, as in this photo taken in Dodge City, Kansas, at 10:00 A.M. on March 30, 1935. (Kansas State Historical Society.)

A black blizzard engulfed the Dust Bowl on April 14, 1935. (Kansas State Historical Society.)

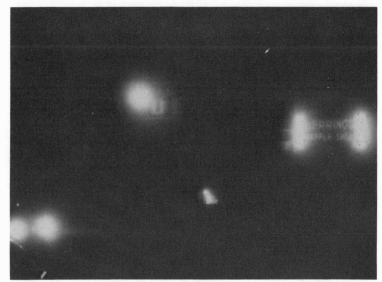

On April 14, 1935, dust clouds dimmed lights to a faint glow at 3:00 P.M. in Dodge City, Kansas. (Kansas State Historical Society.)

Dust storms were common over the southern Great Plains from 1932 to 1940. (Kansas State Historical Society.)

Farming and travel became difficult as dust storms blew drifts of soil along fence rows and roads. (Kansas State Historical Society.)

The ground on either side of this sidewalk in Syracuse, Kansas, was level with the concrete before storms blew dust into town during the 1930s. (Library of Congress.)

Dust Bowl residents often spent Sunday afternoons in rabbit drives. Rabbits were driven into a pen, clubbed, and stacked. (Kansas State Historical Society.)

A typical dust storm in April 1936 brought a haze to the Dust Bowl. (Library of Congress.)

This basin listed field will collect rainfall and retard soil movement. (Library of Congress.)

Furrows near Dalhart, Texas, helped hold blowing soil in June 1938. (Library of Congress.)

A Liberal, Kansas, farmer listed his field in March 1936 to prevent soil blowing. (Library of Congress.)

Drought relief cattle were penned in Kansas City, Kansas, stockyards. (Kansas State Historical Society.)

A Dust Bowl farmer in Cimarron County, Oklahoma, raised a fence that was nearly buried by dust. (Library of Congress.)

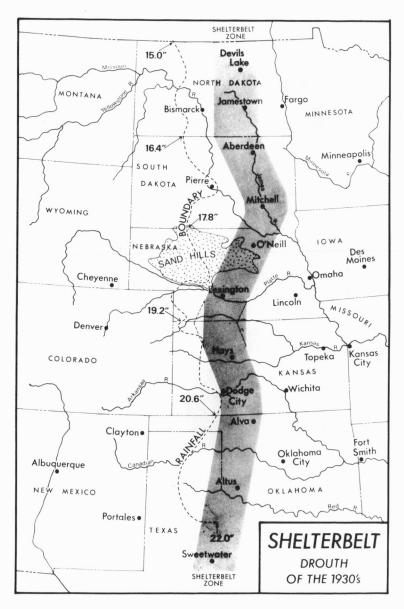

The dark area indicates the shelterbelt zone. (Nebraska State Historical Society.)

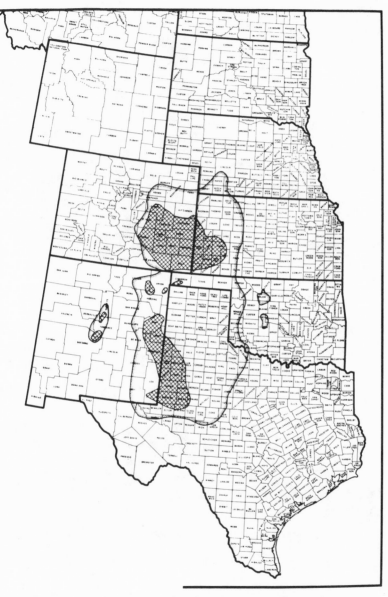

This map shows the general extent of wind erosion up to April 1, 1954. More than 50 percent of the cropland in the checkered areas had been damaged by wind erosion. (Smithsonian Institution.)

Soil drifted heavily on this Oney, Oklahoma, farm in 1950. (National Archives.)

In Hamilton County, Kansas, soil drifted to the top of a five-foot-high fence during the dust storm of February 19, 1954. (National Archives.)

Crops failed in 1950 near Raymondville, Texas. A year later the soil was almost completely bare, and the dust was drifting badly. (National Archives.)

A dust storm on February 19, 1954, created scenes like this in Stanton County, Kansas, in only twenty-four hours. (National Archives.)

A grain drill left in a Stanton County, Kansas, field was nearly covered with soil in the dust storm on February 19, 1954. (National Archives.)

Near Syracuse, Kansas, blowing soil had almost filled Sand Creek by spring 1954. (National Archives.)

CHAPTER SIX

Agricultural Problems and Relief

Compared to the "dirty thirties," southern Great Plains agriculture was quite productive during the 1920s even though export markets were declining and the nation sank into depression in 1929. By 1931, though, agricultural problems became critical when a record wheat crop brought a precipitous collapse in prices. In southwestern Kansas the price of wheat typified the entire problem in the Dust Bowl. There prices dropped to thirty-three cents a bushel in 1931 and to twenty-nine cents the following year. With large machinery and land indebtedness contracted during the previous decade but not paid for, Dust Bowl farmers would have encountered financial difficulties even if high yields had been maintained. But high productivity was not sustained, as drought and dust brought agricultural production almost to a halt.[1]

From 1931 through 1937, the southern Plains experienced a general period of moisture deficiency, and drought became the most severe on record from 1934 through 1937. Because of low prices and increasingly dry seeding conditions, Dust Bowl farmers as a whole planted less wheat in the autumn of 1932 than they had a year earlier. As wheat production declined, farmers planted more corn and grain sorghum. But from 1931 through 1932, corn and hay crops fared little better than wheat by ranging from 50 to 80 percent of normal.

During the following two years corn production dropped to less than 35 percent of normal, and by 1937 farmers had almost completely abandoned it for sorghum because corn took even more moisture than wheat to grow successfully.[2]

Searing drought continued into 1933. With each passing month the chances for a successful wheat harvest diminished. No precipitation fell in Cimarron County, Oklahoma, during March, and the Liberal, Kansas, area received only .25 inch for the same period. By April agricultural experts reported an unusually high percentage of the Kansas wheat crop abandoned because of the drought. Near Elkhart only 37 percent of the crop approached normal, and in Haskell County, Kansas, it only reached three percent. In the Oklahoma and Texas Panhandles similar conditions prevailed. As a result, the Dust Bowl wheat harvest averaged only about two bushels per acre. In short, it was a total failure, yielding little more than the seed sown.[3]

Partly to avoid future disasters such as the 1933 harvest, some Texas Panhandle farmers began to plant cotton as far north as Ochiltree and Lipscomb Counties. Although the initial planting was small—less than 150,000 acres—the editor of the *Amarillo Globe* hoped cotton would help diversify Panhandle agriculture and create a "balanced farm program." Indeed, diversity was needed, since of the 2.5 million acres of wheat planted in the Texas Panhandle only about 800,000 acres were harvested in 1933.[4]

Diversification came slowly, though, as farmers began to make up for poor harvests by planting more wheat. West of Woodward, Oklahoma, most of the land was plowed for wheat by late August 1933. The soil was dry, however, and few farmers maintained hope for adequate seeding conditions that autumn. The Amarillo area was over twelve inches below normal by that time. Still, a modest late summer rain sent northeastern New Mexico and Texas Panhandle farmers into the fields to make a final effort to plant feed crops. The precipitation varied in amounts from .25 to 2.50 inches and prompted a meteorologist at the Amarillo weather bureau to call it the beginning of a "rainy cycle." That judgment was overly optimistic. Although the rain helped feed crops in some localities, it was, as a whole, insufficient to aid germination and the feed crop was severely stunted by the drought.[5]

The drought continued into the autumn of 1933 and hot winds blew wheat seed out of the soil in some areas. By January 1934, the wheat crop for the Texas Panhandle was estimated at only 60 percent of normal. By mid-February, 50 percent of the wheat crop was destroyed by wind and drought in Hartley and Dallam Counties. A month later in Cimarron County, Oklahoma, the county agent estimated that 25 percent of the wheat crop had been destroyed and predicted a harvest of only four bushels per acre, where in normal times farmers harvested twenty-five bushels per acre with little difficulty. Drought conditions were much the same in southwestern Kansas where the wheat crop was down 246,000 acres from the year before. The harvest prospects seemed to improve, however, in March 1934 when a light rain spread over the Texas Panhandle. Agricultural experts almost immediately predicted a 20 million bushel harvest—up 5 million bushels from the January estimate. Dust Bowl farmers were optimistic that the drought had been broken, and they planned for the biggest harvest since 1931. A local editor reflected the general optimism of the region when he wrote "the goose hangs high." Panhandle-Plains dirt farmers, he wrote, were "going down the furrow with brighter days in sight and a song in their hearts."[6]

Bright days and optimism lasted only until a May dust storm obscured the sun and destroyed the wheat crop. At the same time, the drought became the worst on record. In thirty-four Texas Panhandle-Plains counties extending as far south as Lubbock, experts now predicted a three to five bushel per acre wheat harvest instead of thirty-five bushels per acre. By early June, the winter wheat crop was only 52 percent of normal in Texas, 47 percent of normal in Kansas, 40 percent and 21 percent of the average in Colorado and New Mexico respectively. When the binders, threshing machines, and combines left the fields, most Dust Bowl farmers were lucky to have cut three bushels per acre. In Cimarron County, Oklahoma, farmers harvested about the same amount as the seed wheat they planted.[7]

Severe drought continued into the summer of 1934 causing "heavy and irreparable damage" to crops and pasture in Kansas, Oklahoma, and Texas. To help to alleviate the agricultural crisis, the director of the Oklahoma extension service urged farmers to plant turnips and to can vegetables. "The coming winter," he said, "is going to test the mettle of the Oklahoma people." Certainly, the immediate future appeared grim, but Dust Bowlers tried to make the best of things. Some believed that because of the drought and the dust storms "young men and women and boys and girls will be given a demonstration in thrift which will prepare them for meeting other trying periods in years to come." Without a doubt, thrift and perseverance were needed because the farm situation became worse.

Across the Dust Bowl, autumn plantings of winter wheat failed to sprout, and the ground remained without soil-holding cover. Once again, severe soil blowing began in the spring of 1935 and caused an estimated $18 million damage to the Texas Panhandle wheat crop alone. Since that crop was only 41 percent of normal, the loss was particularly distressing for those Texas farmers. In Oklahoma, Harry B. Cordell, president of the state board of agriculture, claimed that April dust storms virtually destroyed the state's wheat crop, and in Kansas, farmers harvested an average of only one bushel per acre. This catastrophic harvest caused one farmer to profess the harvest was so bad that he cut thirty acres to the bushel and that two sparrows trapped in his combine ate the crop faster than he could cut it.[8]

Withering drought continued into 1936, and excessively hot temperatures made the agricultural situation even worse. In Kansas, temperatures ranging from 100 to 120 degrees during the harvest season destroyed much of the wheat crop, which averaged only three bushels per acre. Drought and intense heat burned the pastures and corn crop in the Oklahoma Panhandle; Colorado and New Mexico farmers expected the feed crops to fail unless rain came immediately. In 1937, the last year before scattered rains eased drought conditions, many Dust Bowl farmers averaged only two bushels of wheat per acre.[9]

Overall, Dust Bowl agricultural conditions are well typified in southwestern Kansas. In the years 1933 to 1937, over three-fourths of the seeded acreage in wheat produced an average yield of less than three bushels per acre; two-fifths of that acreage yielded less than one bushel. The corn crop was little better. Two-fifths of the corn acreage failed during that period, and half the acreage produced less than one bushel per acre. Sorghum grains, normally drought resistant, also suffered under the searing heat and persistent drought. This crop averaged less than three bushels per acre in 1935 and 1937. During the two most severe drought years—1934 and 1936—the yields were so small that they were not reported. During the dry years after 1932, farmers made no concerted effort to reduce their wheat acreage. They planted more grain sorghum in 1933 and 1935 when wheat abandonment was high, but they decreased those plantings in 1934, 1936, and 1937 when they harvested about half their wheat crop. Continued wheat failure also hindered diversification by forcing farmers to reduce the number of hogs they were feeding with the excess grain from the bumper wheat crop of 1931. As a result, the number of hogs dropped from 147,000 in the spring of 1932 to 26,000 four years later for the region as a whole.[10]

As Dust Bowl farmers teetered on the verge of failure and bankruptcy, federal aid designed to alleviate the hard times in the farm community caused by the depression was invaluable in helping farmers to stay on the land until the rains returned. Although federal programs provided economic relief from drought and depression, the aid from the Agricultural Adjustment Administration (AAA), created by the "First Hundred Days" Congress in 1933, was the most significant.

Without AAA financial help, Dust Bowl farmers would not only have gone hungry, but the rate of farm abandonment would have been disastrous.[11]

The AAA program had two major phases, the first to provide payment for farmers who reduced production of certain basic commodities such as wheat, corn, cotton, rice, tobacco, and hogs. Dust Bowl farmers received payments primarily for wheat and swine reductions. In 1936, when the Supreme Court held the AAA Act of 1933 unconstitutional and Congress therefore had no power to pay farmers to follow a production control program, Congress quickly responded with the Soil Conservation and Domestic Allotment Act. That legislation was designed to insure continuity of the farm relief program and, by so doing, began the second phase of the AAA aid to farmers. Under the latter law, farmers still received payments designed to limit production and increase purchasing power, but they were paid instead for not planting soil-depleting crops and for sowing soil-conserving grasses, legumes, and feed crops. The shift of emphasis from production controls to soil conservation provided more equitable aid to cooperating farmers. Dust Bowl farmers particularly benefited from the second AAA program because they could earn payments on any part of their crop lands which contributed to soil conservation instead of on only a few select crops designed to limit production.[12]

Nearly all Dust Bowl farmers participated in the AAA program. Indeed, they could not afford to do otherwise. Although they would have preferred to plant as much wheat as possible and hope for rain, they had little choice but to accept AAA checks. They needed the money. The only significant opposition to this New Deal program came from the large grain dealers, commission men, and food processors (who were taxed to pay for the first AAA program), rather than from the farmers. As a result, AAA payments became the major form of income for many farmers between 1933 and 1937.[13]

Payments varied directly with the number of acres in crops and with the total farm acreage operated. Since payments during the first phase of the AAA were based on reduced production of basic commodities—in this case wheat—payments were higher to larger farmers who had more acreage in wheat. During the first six months of 1934, Baca County, Colorado, farmers received more than $190,000 in wheat production control or allotment checks. In Hodgeman and Meade Counties in Kansas, the AAA payments for that same period were more than $286,000 and $333,000 respectively. Cimarron County, Oklahoma, farmers received a substantial amount of federal aid in this form as well, since AAA checks totalled nearly $363,000 for that time. In a twenty-eight county area of the Texas Panhandle-Plains, farmers had received nearly $7.5 million in AAA checks by the end of 1934.[14]

Dust Bowl farmers also welcomed the AAA's corn-hog reduction program because it also provided a chance to gain additional income. That program was designed to reduce nationwide corn acreage by 20 percent and hog tonnage by 25 percent. Although Dust Bowl farmers raised little corn and few hogs, the program was beneficial to participants. For example, during the two year period ending in March 1935, corn-hog checks totalled nearly $28,000 for Meade County, Kansas, and $58,000 for Cimarron County, Oklahoma, and from July through September, farmers in forty-six Texas Panhandle-Plains counties received more than $278,486 for complying with this aspect of the AAA program.[15]

In the spring of 1937, the AAA announced an emergency wind erosion control program which provided additional economic relief for the drought stricken area. Under this program, farmers received financial aid for the adoption of certain wind erosion control practices. This aid was in addition to the payments received for diverting acreage from production.

For example, a farmer could divert 15 percent of his acreage from production and receive a AAA payment for doing
so. Additional acreage still subject to wind erosion was eligible for an allowance of seventy-five cents per acre for financ-
ing soil conservation tillage measures. In other words, a Dust
Bowl farmer with 300 acres in crops could receive AAA payments in the following manner. First, the AAA would deduct
15 percent which was the amount of cropland diverted for
payment under the general program. This would leave 255
acres of which the AAA might still consider 160 acres subject
to wind erosion. The allowance for the wind erosion control
program would be 160 acres times seventy-five cents for a
total allowance of $120. Under this aspect of the AAA's wind
erosion control program, Dust Bowl farmers could earn
twenty-five cents per acre for contour listing. Based on the
above example the payment would be $40. If the farmer
planted a sorghum grain cover crop as well, he earned an
additional fifty cents per acre or $80 for a grand total of $120
for soil conservation.[16]

Dust Bowl farmers also received financial aid from the
Resettlement Administration (RA) which Franklin Delano
Roosevelt created in April 1935 to help with the problem of
rural poverty. Only those farmers who had exhausted all
other forms of credit were eligible to apply for RA aid in the
form of "rehabilitative" loans. These loans allowed farmers to
purchase necessities—food, clothing, feed, seed, and fertilizer—for the purpose of making the farm operator self-suffi-
cient once the drought ended. Before making a loan, the RA
designed a farm management program which budgeted the
farmer's expenses so that the operator could know how much
to spend and still meet other loan and mortgage obligations.
Resettlement Administration loans made in the Dust Bowl
averaged about $700 per farm family, and, by mid-August
1936, these loans totalled $30,000 for the Texas Panhandle,
$50,000 for southeastern Colorado, $35,000 for southwestern Kansas, $47,000 for the Oklahoma Panhandle, and
$109,000 for New Mexico. One Dust Bowl editor believed

the RA loans would encourage farming diversity, particularly the use of multipurpose cattle, and had gotten "the ball started rolling toward big hog and dairy herds in the country like in the old days when the people lived on cream and egg checks and blew their wheat checks on extras." Although few Dust Bowl farmers had ever lived that way, these loans were of substantial benefit in helping farmers cope with dust, drought, and depression, and by so doing enabled them to maintain their operations until the rains and good times returned.[17]

The Farm Security Administration (FSA) which replaced the RA in 1937 also made loans to farmers who could not obtain credit from other sources. These so-called "standard loans" were granted to farmers whose operations promised to be self-sustaining provided they had adequate access to equipment, seed, and livestock. The FSA loans were also designed to enable farmers to shift emphasis from cash grain farming to mixed farming with greater emphasis on raising livestock. Dust Bowl farmers then leased additional lands in order to establish an operation that would be large enough to assure success even during the most severe droughts. Because such change would take a long time, these loans were repayable over a ten-year period. It was FSA strategy to provide such loans to Dust Bowl farmers in hopes they would use this money for renting land from nonresident owners, and thereby prevent wind erosion if the land started blowing.[18]

Despite aid from the AAA, RA, and FSA, Dust Bowl farmers and town dwellers alike had a great difficulty meeting their financial obligations, and tax delinquencies increased. In 1929, for example, only 2.9 percent of the taxes in Roberts County, Texas, were delinquent, but in 1932 more than 43 percent of the property taxpayers could not pay. Farther to the north in Cheyenne County, Colorado, tax delinquencies rose from 1.4 percent in 1930 to more than 24 percent in 1933. Farmers were particularly unable to pay their taxes because in some areas 90 percent of the farm families were on relief. In the Oklahoma Panhandle, for example, by late

summer 1934, scarcely more than half of the taxes had been
paid in Cimarron County. A year later more than 58 percent
of the farm land in Childress, Deaf Smith, Randall, and
Swisher Counties in Texas was tax delinquent. The farm situation worsened as property valuation declined up to 90 percent in some areas. As farm valuation shrank local tax
revenues, higher property taxes were imposed. In Baca
County, Colorado, for example, land was valued at more than
$10 million in 1920 and taxed at the rate of 6.75 mills. By
1936 the valuation had dropped to slightly more than $8 mil-
lion, the levy had risen to 9.5 mills, and about 30 percent of
those taxes were delinquent.[19]

As wheat prices and property values fell, and as property
tax rates rose, tenancy and nonresident land ownership
increased from 38.5 percent of all operators in 1935 to more
than 42 percent in some counties five years later. Nonresidents in 1940 owned nearly half of the farmland in Dallam
County, Texas. In Hale County, Texas, and Curry County,
New Mexico, the increases were from 40 to 55 percent and
from 18 to 39 percent respectively. Tenants on land owned
by nonresidents relied heavily on federal aid. In 1934,
tenants operated 44 percent of the land under corn-hog contracts in Dallam County and 52 percent of the land under cotton contracts in Hale County. The following year, tenants
comprised approximately 75 percent of the relief clients in
Dallam and Curry Counties.[20]

The economic plight of Dust Bowl farmers became even
more serious as mortgages increased. In 1934, 850 Cimarron
County, Oklahoma, farms bore approximately $4.75 million
in mortgage indebtedness for an average of more than $5,500
per farm. Of that total, $1 million was for chattel mortgages,
and the remainder was for real estate. Oklahoma law permitted a chattel mortgage to be filed against crops not yet
planted. Consequently, it was common for chattel mortgages
to be imposed against crops planted three to five years later.
Oklahoma farmers complained that this practice meant that a
one-year lien often became a three to five-year mortgage. In

1929, Cimarron County farmers purchased more than $896,000 in machinery, largely on credit, but few farmers had made payments on it by 1934. In that year, their indebtedness was so great that it would have taken the entire income from every wheat crop since 1932 to have paid the chattel mortgage indebtedness there alone. Clearly, these farmers were in serious financial trouble, and the implement companies began sending agents door to door to persuade farmers to assign their AAA checks over to them. One agent estimated that his company received seven out of every ten checks asked for—50 percent of which were coerced under threat of foreclosure.

Most county agents and others were usually able to intervene and gave the implement companies so much bad publicity when this happened that they withdrew their demands. But bankers were also guilty of pressuring farmers to turn over their AAA checks. In Randall County, Texas, for example, where the farm debt averaged $6,710, one banker urged a farmer to apply his AAA wheat check to the interest on his loan. This, he said, would keep the farmer in "good standing" in case he needed additional loans later. County officials also sometimes asked farmers to apply their AAA checks to their taxes, and one farmer received a letter from a piano company thanking him in advance for remitting as much money as possible from his AAA checks.[21]

Although wheat prices fell because of overproduction and although drought and dust storms ruined crops and caused additional economic hardship, farmers did not exodus in mass from the Dust Bowl. Steinbeck was wrong; the Okies and others like them were not products of the Dust Bowl, but of the cotton region east of the most drought-stricken areas. Certainly, *The Grapes of Wrath* is an important social commentary on the plight of the dispossessed in California during the Great Depression. Many families like the Joad's did flee the Dust Bowl over Route 66, but most of them did not leave. Most California migrants were products of the drought across the entire Great Plains and Midwest and not just from the

Dust Bowl. Only three counties in the Oklahoma Panhandle were located in the Dust Bowl, and from 1930 to 1940 the total loss in population there was only 8,762. Many of the Panhandle farmers moved, but not to California. Instead, they fled to the nearest town where they could be closer to the employment and relief offices.[22]

This is not to say that many Okies did not enter California during the drought years, because they did. Of the 43,180 migrants in need of employment who entered the state between June 16 and December 15, 1935, over 7,000 or 16.5 percent came from Oklahoma. This number was double the migrant total from the second ranking state—Texas. More Oklahoma migrants followed. From July 1935 through June 1939, 22.7 percent of the emigrants in need of manual employment in California were from the Sooner State, but most of these Okies were from the cotton region where the economy had totally collapsed. There farmers produced more than 1.5 million bales of cotton at $22.33 per bale in 1924, but only 854,000 bales at $8.71 per bale in 1930. A year later, the price of cotton dropped still further to $5.06 per bale, and prices continued to drop through 1935. Furthermore, Oklahoma cotton was short-staple cotton which was not in high demand on the international market. Thus, with little international demand and with the drought shriveling the crops in the fields, many cotton farmers quickly reached economic destitution. In addition, while most wheat farmers owned their own land, more than 60 percent of the Oklahoma cotton farmers were tenants or sharecroppers. Because of low prices and decreased harvests, many landowners in the cotton producing region preferred to release their tenants and cut back on production in order to collect AAA allotment checks. When they did so, thousands of tenants and migratory farm workers were forced onto the highways in search of employment elsewhere. Displaced cotton tenants and field workers, then, were the Okies, not the Dust Bowl wheat farmers.[23]

The migration situation was much the same in the Texas Panhandle, where twenty-three counties lost fewer than 15,000 inhabitants between 1930 and 1940. Nine counties actually gained population because of the booming gas and oil field industry. In a twenty-seven county area in southwestern Kansas, the number of farmers increased from 22,369 in 1930 to 23,916 in 1935. Some farmers left, but they were the ones who had little money invested in their lands. Most farmers who remained in the Dust Bowl had too much invested to leave, and the government helped them to stay.

Although wheat was the predominant crop, relatively few farmers relied entirely on it. Most raised some livestock and planted feed grains. Even if diversification was minimal, wheat, livestock and feed grains, plus government aid, enabled them to retain their holdings and remain on the land until the drought ended. Even so, life was difficult at best for Dust Bowl farmers. For example, one Wheeler County, Texas, farmer who was refused credit at a local grocery store picked up a sack of flour and went home. When the county sheriff arrived, he found the farmer's wife mixing flour, salt and water for dough and the children eating it before it was baked. With conditions similar to this, though perhaps not always quite as severe, all across the Dust Bowl, government aid alone could not restore the region to prosperity, only rain could do that.[24]

While Dust Bowl farmers waited for the rains to return, they expanded their operations, largely because of government aid. In the Kansas portion of the Dust Bowl, for example, the total acreage in farms rose from 73.4 percent of the land area in 1930 to 78.4 percent by 1935. Crop acreages also increased from 292.2 to 344.3 acres per farm or from 26.4 percent to 33.2 percent of the total land area. Even so, these farmers harvested less than half as much cereal grain in 1935 as they had five years earlier. Generally, the larger farmers had a higher percentage of land in crops than did the smaller farmers, and the larger farmers continued to expand their crop acreage at the expense of their pasture lands while the opposite was true of the smaller farmers. Few small operators actually reseeded cropland to grass. Instead, the change in land use resulted from the movement of many small farmers to lands with a high proportion of acreage still in sod, and by the purchase by the larger farmers of lands already broken for wheat.[25]

Agricultural conditions began to improve in the spring of 1938 as precipitation increased and as the dust storms diminished in number and intensity. In southeastern Colorado, subsoil moisture was the best since 1931. Across the Dust Bowl, the wheat harvest that year was fair, but compared with the past few years it seemed tremendous. In Beaver County, Oklahoma, farmers harvested 264,700 acres (9,500 more than in 1937) and reaped more than ten bushels per acre. In fact, the wheat harvest was so large that a temporary labor shortage prevailed in some areas, and the mayor of Hooker, Oklahoma, urged local businessmen to release their employees to help with the harvest. Other Dust Bowl farmers had comparatively large harvests as well. In southwestern Kansas wheat production doubled in most counties. Several months later, in mid-September, one Dust Bowl newspaper reported: "Baca County has made a miraculous recovery and its greatest need just now is an influx of new manpower and new settlers to work the soil and resume development of the county where the dust forced its cessation." By mid-December, farmers were once again carelessly burning the soil-holding thistles from their fields and along the roads. One local editor suspected that these "flaming ghosts" would haunt them during the spring blow months, if they did not stop that practice immediately.[26]

Additional precipitation early in 1939 ended the dust menace and agriculture began to boom once again. By mid-June 1939, agricultural experts were predicting a wheat harvest of 15 million to 25 million bushels in the Texas Panhandle—over six times the 1935 harvest. That optimism was warranted and the harvest was the best in seven years, which caused one editor to proclaim that the Dust Bowl had become a "horn of plenty." With the drought broken and the dust settled, land that could not be sold for $12 an acre in 1935 now brought as much as $35 an acre in some localities. War in Europe expanded the wheat market and raised prices. Consequently, a land-leasing boom began as residents and nonresident farmers tried to capitalize on the improved agricultural conditions. Early in 1940, nearly seventy quarter-sections of land changed hands in Baca County, Colorado, within a few weeks time. With the return of normal precipitation through most of the next decade, Dust Bowl farmers quickly forgot or chose to overlook the past. That negligence was to cause a return of the wind erosion menace during the 1950s.[27]

In retrospect, even in the most fruitful of years, the critical margin between success and failure in the Dust Bowl is so small that only the best of farmers have any chance of doing well. During the 1930s, though, even good farm management techniques were inadequate to prevent economic hardship when farmers faced the triple problems of drought, dust, and depression. Certainly, government aid in various forms from a multiplicity of agencies gave farmers buying and refinancing power which they would not have had otherwise. No farmer got rich from government aid, but without it the Dust Bowl agricultural community would have suffered far more even than it did. Still, while farmers tried to contend with the problems of blowing soil and withering crops, they were faced with another agricultural crisis—dying cattle.

CHAPTER SEVEN

The Drought
Cattle Emergency

In Biblical times, the prophet Joel wrote, "The herds are perplexed, because they have no pasture. . . ." Much later, during the 1930s, Dust Bowl cattle were more than perplexed—they were dying from starvation and suffocation. Drought, heat, dust, and overgrazing were causing serious concern among cattle producers by the summer of 1933. In August, the pastures were in "poor to very poor" condition, and, although there were fewer cattle in the Dust Bowl than earlier in the twentieth century, cattlemen nationwide had overproduced and the price was dropping. Between 1928 and 1934 cattlemen increased production nearly 20 percent and the number of cattle sold for slaughter rose by 3 million head. That production was enough to supply a population increase of about 20 million based on the current rate of beef consumption. As nationwide beef production increased, prices fell.

During 1933, beef prices dropped to the lowest level since 1899—$3.63 per hundredweight. That price was more than $2.00 below the fair exchange value based on pre-World War I costs, and showed a reduction of at least $5.50 from the average price paid in 1929. Since cattlemen had fixed debts and other expenditures, these low prices meant economic disaster. When the drought and dust storms sharply reduced the feed supply, the Dust Bowl's livestock industry became even more tenuous than it had been a few years earlier.[1]

Early in 1933, Congress provided some relief for Dust Bowl cattlemen by authorizing federal loans for the purchase of livestock feed. Cattle producers could apply through the Farm Credit Administration (FCA) and receive $2.50 per month for each head of horses or cattle over one year old, thirty cents per head of sheep, and a dollar for each brood sow. These FCA loans could not exceed four months or $250, and they were due on or before August 31, 1934. Since the FCA feed loans carried an annual 5 percent discount rate which was deducted in advance, this aid was greatly reduced in value from the very beginning. Furthermore, the FCA took a first chattel mortgage on all livestock for which the feed loan was granted. If a prior chattel mortgage existed, the farmer had to obtain waivers in favor of the FCA before receiving the money. By late August 1935, nearly twenty thousand borrowers in the five Dust Bowl states had received over $6 million in emergency feed loans.[2]

In Kansas, Governor Alfred M. Landon appealed to the national office of the Red Cross for additional aid. On June 17, 1933, Landon wrote, "In some counties help [is] absolutely necessary within twenty-four hours in securing feed for livestock and work stock. . . . Farmers have generally exhausted all credit and must have help through [the] Red Cross." Landon insisted that the Red Cross act quickly or else announce its inability to cope with the drought problem. The Kansas governor also informed Harry L. Hopkins (head of the Federal Emergency Relief Administration) that the Dust Bowl needed cow feed rather than sympathy and federal aid rather than red tape to end the drought crisis.[3]

Santa Fe Railroad officials recognized that if the Dust Bowl livestock industry failed, the company would suffer serious financial losses due to decreased shipping opportunities. Since the railroad received a large amount of its business from shipping livestock out of the Dust Bowl and from hauling needed cattle and other agricultural supplies into the area, it responded quickly to the governor's appeal for aid. It did so by declaring "emergency rates" which reduced the cost for shipping carload lots of hay and straw by 50 percent. Poultry and livestock feed was reduced to two-thirds, and cattlemen were charged only 85 percent of the normal rate for shipping livestock to pastures elsewhere. Even with these measures, however, the feed situation remained critical across the Dust Bowl. By late August 1933, Oklahoma feed grain had declined by 1.6 million tons compared to a year earlier, and Governor Landon called for "extreme measures to conserve" and urged cattlemen to make use of every ounce of feed available.[4]

Because of the drought conditions and poor feed crops in the summer and autumn of 1933, livestock came through the winter in poor condition. Farm Credit Administration feed loans and reduced freight rates were insufficient to meet the crisis, but cattlemen tried to retain their livestock as long as possible while they waited for rain and the return of good pastures.[5]

Dust Bowl cattlemen needed more than just a return of the rain to revitalize their industry. Actually, additional aid was available through the AAA, at least in the form of protection from overproduction. The original AAA act in 1933 included cattle as a basic commodity for regulation, but that part of the bill failed in the Senate after passing the House. Dust Bowl cattle producers opposed production limits because they believed it would force consumers to eat less beef. Instead, they preferred higher tariffs, lower interest rates, and marketing help. By spring 1934, however, continued drought, economic distress, and administration urging convinced cattlemen to accept a production control program. Before it could be implemented, though, the drought intensified and forced the federal government to take emergency action to save the cattle industry.[6]

Indeed, the drought reached appalling severity. In New Mexico, for example, farmers abandoned nearly all of the wheat crop. Ranges and pastures offered only the scantiest amount of grass, and all dry feed was grazed close to the ground. By June 1, 1934, a large number of cows and lambs were dead. Cattlemen were feeding concentrates and moving their herds to better ranges. Unless relief came soon, they believed the livestock death rate would be "appalling." Similar situations existed in Kansas and Colorado; all prospects for corn were "gloomy."[7]

Dust Bowl cattlemen received some of the help they wanted on November 10, 1933, when the Federal Surplus Relief Corporation (FSRC) began purchasing beef as a price support measure. In April 1934, the Jones-Connally Amendment to the Agricultural Adjustment Act also provided funds for cattle price supports by making livestock a basic commodity. Still, the government buying program was on a limited scale. In mid-May 1934, the federal government announced a major livestock relief program with the creation of the Drought Relief Service (DRS) to assume responsibility for all AAA and USDA relief work and to coordinate it with the activities of the Farm Credit Administration and Federal Emergency Relief Administration. Most importantly, the DRS was to buy cattle in designated emergency counties within the drought area. However, cattlemen had to have clear ownership of the animals sold, or else the lien holder had to agree to the sale. These provisions were designed to aid cattle producers suffering economic distress and to avoid speculation in the cattle market. Emergency counties were those where the drought was particularly acute and where livestock were threatened with starvation. In secondary drought counties, cattle producers could receive reduced freight rates for feed and cattle shipments, but they could not participate in the purchase program. By the end of the summer, though, almost every county in the Dust Bowl had been declared an emergency area.[8]

The actual buying operation was directed by the commodities purchase section of the AAA. The prices paid depended on the age and condition of the cattle. Payments were divided into two categories—benefit and purchase. The owner or lien holder was to receive the purchase payment, and it could be used to satisfy the mortgage. The benefit payment belonged to the producer alone, but it obligated him to support a future production control program. The amount paid depended on the age of the cattle, and the prices ranged from $3 to $6 for benefit payments, and $1 to $14 for purchase payments. Local agents, usually appointed by the county agent, appraised the cattle. At first the appraisers visited individual farms. Later, in order to speed the purchase program, farmers took their cattle to designated railway shipping points on specific days and the appraisal was made there. An agent from the Bureau of Animal Husbandry accompanied the appraisers and condemned all animals not suitable for human consumption.[9]

The first DRS purchases of Dust Bowl cattle came in early June 1934. These cattle were handled in two ways: first, all animals unfit for food were condemned and destroyed at the point of purchase; second, all cattle other than foundation stock were purchased and donated to the Federal Surplus Relief Corporation for slaughtering and distribution to relief families nationwide.[10]

The Emergency Cattle Purchase Program became effective with unprecedented speed, and many Dust Bowl cattle producers were able to sell a substantial number of livestock before the cattle died from heat, starvation, or suffocation. Many cattle immediately offered for sale were "culls"—cattle entirely unfit for shipment and slaughter. Consequently, at first, the condemnation rate was high, often exceeding 50 percent in some localities. A large number of calves were condemned because they were stunted from insufficient feed and water and could not withstand long-distance shipping and still arrive fit for slaughter. Some calves were used for local relief purposes, but most were destroyed and buried.[11]

All state directors of the Emergency Cattle Purchase Program were to allow purchases only to the extent necessary to insure maintenance of foundation herds during the coming winter. Actually, government prices (which were lower than the private market rates) did not encourage individual producers to sell any stock that could be held for sale later on the open market; rather they stimulated the selling of the unfit. For that reason, the DRS did not make a distinction in prices for different grades, and Dust Bowl farmers were inclined to retain only their best animals, since feed was in short supply and expensive to buy.[12]

Emergency drought cattle purchases were to terminate on September 12, 1934 when the funds allocated for the program would be exhausted. But, the continuation of the drought and the poor prospects for a good feed crop brought additional federal money for the program. All livestock purchases were to close definitely on February 1, 1935. In Kansas, though, the termination date was extended to February 15, which greatly benefited Dust Bowl cattlemen since feed supplies were at least 50 percent below normal.[13]

Not all cattle delivered to the FSRC were shipped directly to the slaughterhouse. Many head were sent first to pastures in the South and East. There, they were fattened and held for later packing. This also helped prevent overburdening the canning plants. Once slaughtered, fresh canned beef was distributed through state and county units of the Federal Emergency Relief Administration or by the FSRC.[14]

Packing plants that had contracts with the FSRC received the drought cattle. Consequently, the Emergency Cattle Purchase Program helped to stimulate this sector of private enterprise as well. The beef canned for the FSRC could not be sold and all cans had to be so labeled. This labeling enabled FSRC beef to be kept separate from that prepared for commercial sale. Some cattle were also processed solely for distribution within the state of purchase, in order to provide work relief. That policy was merely an extension of the procedure which the Texas Relief Commission (TRC) developed in late 1933. At that time, the TRC had received a grant from the FSRC to buy cattle and establish canning plants. When the project ended in January 1934, ample precedent had been established for a similar program at a later time. By establishing state canning plants, the FSRC linked the cattle purchase program, which benefited the producer, with a work relief program which helped the unemployed.[15]

In Kansas, the FSRC appointed the Kansas Emergency Relief Committee (KERC) to receive, purchase, and ship cattle. The KERC had the authority to purchase as many animals as it could use in the state program. Animals too weak to ship or feed for later slaughter were destroyed, but cattle strong enough for shipment went immediately to the packers for canning. Some, however, were pastured in southeastern Kansas for later processing. Twenty days after the DRS cattle purchase program began in Meade County, Kansas, the federal government had purchased 11,421 head, an average of 557 cattle per day. Because even more cattle were to be purchased in Kansas daily, the KERC signed contracts with private feedlots to care for the animals until arrangements for packing could be made. In less than three weeks, 9,000 head were being held in those feedlots and an additional 2,500 were on southeastern Kansas pastures.[16]

At first, most of the drought cattle were shipped out of the state, but to augment the Kansas relief purchase program, the KERC received permission from the FERA to set up its own canning plants at various locations within the state. The first plant began operation at Parsons in mid-August. Eight other "kitchens" were established in Kansas City, Wichita, Topeka, Leavenworth, Hutchinson, Coffeyville, Independence, and Chanute, and about 9,000 relief workers were employed. The cattle slaughtered provided fresh beef for the state relief program and KERC cattle supplemented the Federal Surplus Relief Corporation shipping program. Thus, in addition to providing work relief, the Kansas canning program helped remove additional cattle from the drought area and gave cattle producers more economic relief.[17]

Private meat packers received cattle without cost. The slaughtering company retained the bone, shank meat, excess fat, and offal as compensation for slaughtering, quartering, chilling, and delivering the carcasses to the canning plants.

The KERC also reimbursed the slaughterhouse for any feed utilized. Altogether, eighteen canning projects in nine counties slaughtered 102,231 head and processed 12,979,382 fifty- to one-hundred-pound cans of beef. A boning project was also initiated in Emporia as a work relief measure. That project processed cattle which were received too quickly for immediate canning. There, the beef was frozen in one-hundred-pound blocks and kept in cold storage until the fresh beef supply had been exhausted in relief distribution, then it was sent to the canneries.[18]

Although the Emergency Cattle Purchase Program had mixed results and support, it provided several important benefits. First, it saved thousands of cattle from starvation and waste. Second, it gave cattle producers the opportunity to cull inferior and surplus animals from their herds and thereby to upgrade foundation stock. Third, the program prevented financial ruin. It enabled Dust Bowl cattle producers to stay in business, improve their herds, and maintain credit. In addition, the Emergency Cattle Purchase Program boosted the morale of local bankers and merchants and increased cattle prices to a profitable level. If the federal government had not purchased these cattle, the market would have suffered even worse depression as the producers flooded it with drought cattle. Indeed, in spite of the purchase program, thousands of cattle were still sold for low prices at the terminal markets. The only significant criticism of the program from cattlemen concerned prices paid, speed of payment, and government failure to purchase more cattle.[19]

Certainly, the economic benefit to cattle producers was significant. In Kansas, Dust Bowl cattlemen sold 316,149 head for $4,529,347. Colorado producers sold 116,589 head for $1,683,219. In the New Mexico Dust Bowl counties, the federal government purchased 173,916 head for $2,278,259. Oklahoma cattlemen sold 84,244 head for $1,128,395, and Texas producers sold 305,723 head for $3,541,240. As a result of the Emergency Cattle Purchase Program, the editor of the *Amarillo Sunday News and Globe* wrote: "Happy days are coming back to the West." Although that view was overly optimistic, the program did prevent complete ruin of the Dust Bowl cattle industry.[20]

As the Emergency Cattle Purchase Program was being activated, in June 1934 the railroads serving the Dust Bowl once again helped to relieve the feed shortage by reducing rates. Those reductions reached 50 percent on hay and feed shipped into the region as well as on cattle shipped out, and they were to be effective until June 1, 1935 or the end of the emergency. Because of the abnormally dry winter and the distressing black blizzards during the spring, however, the railroads extended the emergency rate reductions for livestock shipped out until July 20, 1935, and shippers could return their cattle at 15 percent the normal rate until June 30, 1936.[21]

These rate reductions failed to improve the feed situation. In August, Governor Landon wrote the FERA concerning the Dust Bowl portion of the state, saying that cattlemen found it impossible to locate adequate feed and water supplies. He urged a doubling of the quota for the governmental cattle purchase program and speedy shipments for the Dust Bowl. In Kansas alone during 1934, an estimated 1 million cattle could not be maintained by the traditional feed sources. If the drought continued, those animals could not be fed through the winter. Since the buying program took 6.000 head per day from the state, only 50,000 head could be removed by November 1, which was considered the end of the pasture season. Dean H. Umberger, director of the Division of Extension at the Kansas Agricultural College, warned Dust Bowl cattlemen not to expect the federal government to buy all their surplus animals. Five hundred thousand cattle would have to be sold through the open market or provided with feed for the winter. Governor Landon believed that the only alternative was for the federal government to increase the state's shipping quota by 25,000 head per day.[22]

Indeed, the feed situation was becoming critical. During the summer of 1934, Dust Bowl pastures furnished almost no feed. In Kansas, feed grain production dropped to 13 percent of the average. The other Dust Bowl states fared little better. Colorado's feed grain production fell to 27 percent, New Mexico to 31 percent, and Texas to 54 percent of normal. Feed reserves left from the past winter were almost nonexistent. Kansas cattlemen were short 78,000 tons of forage and feed grains. Oklahoma producers were in need of 170,000 tons. In Texas and New Mexico cattlemen needed 462,000 and 14,000 tons respectively. Because of the feed shortage, some cattlemen began to sell their herds, and major runs of distress cattle were made on the Kansas City and Wichita markets in early June. Stock water also became scarce, particularly in eastern Colorado where rainfall was 45 percent below normal. There the drought was critical, and cattlemen were hauling water in order to maintain their herds. In Oklahoma, cattlemen often moved their herds four to eight miles a day for water, far more than the livestock could manage and still maintain good physical condition.[23]

By late September 1934, the Russian thistle was a basic livestock ration in the Texas Panhandle. Although the drought had destroyed most of the row crops, the thistles thrived. If thistle hay could be cut before it became woody, it made a tolerable forage when salted and supplemented with cottonseed meal. In Kansas, the state's Emergency Relief Committee established guidelines for the sale of thistle hay from state lands. Although federal inspection was not required to insure quality, thistle hay was to be sold only to dealers certified by county drought relief committees. These committees were to appraise its value, but the price was not to be less than $3.00 per ton of loose hay, $3.75 per ton for baled hay, or forty cents per hundred pounds for ground or stacked hay. The money received went to the Kansas Homestead Rehabilitation Corporation. Tests at the Ft. Hays experiment station indicated that good quality thistle hay compared favorably with hay from other crops and that it had about the same feeding value as wheat straw. This finding was fortunate, because in 1934 thistle hay was about the only cattle feed producers had available. In Kansas more thistle hay was harvested than ever before, and, by early March 1935, the KERC had sold seventy-seven tons baled by relief labor for an average price of $2.75 per ton.[24]

Dust Bowl cattlemen also began using soap weed or yucca for ensilage late in 1934. According to Harley A. Daniel, acting director of the Goodwell experiment station, yucca contained about the same amount of minerals as legumes and provided a satisfactory cattle feed when mixed with grain and cottonseed meal. Texas Panhandle and New Mexico cattlemen began feeding a mixture of yucca, cottonseed meal, and syrup early in 1935 at a cost of only $3.50 per ton. By spring, even that meager feed supply was insufficient to meet the cattlemen's needs. Late May and early June rains throughout the Dust Bowl allowed the weeds to grow, particularly the Russian thistle, but only irrigated hay was growing, and by August the sorghum crop appeared lost even if more rain came.[25]

Because of the severe feed shortage, the Rural Rehabilitation Corporation (RRC) began cooperating with the Department of Agriculture to make its facilities available for supplying feed where shortages occurred or where dealers were unable to keep adequate stocks on hand. The RRC took feed orders from farmers who had emergency feed loans and from others unable to procure forage. Furthermore, since credit for cattlemen was difficult to obtain, the RRC developed a plan to guarantee reasonably priced feed until the drought emergency passed. In order to do so, the RRC made arrangements with state relief administrators whereby feed was allocated to county relief administrators. These individuals in turn supplied the RRC feed to all cattle producers who obtained certificates from the county agent. Those certificates testified that the cattleman owned the stock and was entitled to federal help during the emergency.[26]

The AAA also granted feed and forage loans at the rate of $2.00 to $4.00 on horses, $1.50 to $3.00 on cattle, and $1.00 for an acre of forage production. These loans did not require security, but, if the cattle were mortgaged, the producer had to agree not to transfer title until January 1, 1936. Non-emergency livestock feed loans did require a first lien by the federal government. By mid-March 1935, the drought was so severe that the federal government waived the limits on loans and direct grants for feed purchases. As a result, by the end of the year, government feed loans totalled $154,186 in Kansas; $145,980 in Texas; $141,880 in Colorado; and $57,000 in Oklahoma.[27]

Even with the cattle purchase program, additional freight rate reductions, and the efforts of the RRC and AAA to locate feed or make loans, the cattle situation remained "very serious." Although many cattle had been moved out of the Dust Bowl for pastures in other sections of the country or sold, the number of cattle retained early in 1935 still exceeded feed supplies. By the end of March, practically all locally raised feed was exhausted, and cattlemen were depending heavily on roughages and concentrates to keep their stock alive. Many Dust Bowl cattle were too weak to ship, especially cows with calves, and good pasture within a reasonable shipping distance did not exist. In western Kansas, the pastures were 35 percent below normal. In the Texas and Oklahoma Panhandles, the winter wheat crop was dead and could not be used for grazing; pasture lands were drifting over. In Texas County, Oklahoma, for example, 10,000 head of cattle were starving; 2,500 tons of grain and 4,000 tons of roughage were needed to sustain them for another thirty days. In nearby Beaver County, 60 tons of feed and 4,000 bushels of grain were needed per week to prevent massive livestock starvation. The situation was much the same in Colorado and New Mexico.[28]

The feed situation remained serious as the drought continued into 1936. In early August, the railroads serving Kansas once again reduced freight rates on hay and other feeds shipped into designated drought counties. Governor Landon complained, however, that these reduced drought rates were of little or no value because the reductions applied only to the feed shipped over a single railroad. Because most feed supplies traveled over several railroad lines before reaching the cattleman, those reductions were not of great benefit. Consequently Governor Landon sent a telegram to the major railroad companies and urged rate reductions on feeds shipped in over several lines, noting, "Saving foundation herds means saving future shippers for the railroads." The companies agreed and the rates were reduced accordingly. Additional railroad rate reductions were granted periodically as the drought warranted until the end of April 1938 and provided substantial savings for drought-stricken cattle producers.[29]

In 1937, additional aid came from the Resettlement Administration, which granted special feed loans, and from the Federal Surplus Relief Corporation, which made grain reserves available to Dust Bowl cattlemen. In Kansas, the FSRC provided 350,000 bushels of yellow corn to local feed dealers in the distress area. By early August, pasture conditions were worse than ever before. In Beaver County, Oklahoma, at least 80 percent of the cattle producers could not raise enough feed to carry even subsistence livestock through the winter. Most farmers were depending on weeds and sudan grass for pasture. When the weeds dried up, hundreds of cattle died, and many farmers were forced to dispose of all their livestock. In order to encourage cattle producers to dispose of as many head as possible, the Resettlement Administration began accepting feed loan applications only when cattlemen pledged to destroy all nonproducing animals. In that way, the RA hoped that the best cattle would be kept through the winter, thereby reducing feed demands.[30]

In order to give additional, though long term aid to Dust Bowl cattle producers, the USDA asked Congress in 1937 for $10 million for the purchase of approximately 2 million acres in the Dust Bowl. Furthermore, the Department asked for $76,000 for the development of new drought-resistant grasses for reseeding projects. M. L. Wilson, assistant secretary of agriculture, appearing before the House Appropriations Committee pledged that these funds would restore and conserve grazing lands that would eventually be leased back to cattlemen. Congress approved the plan, and the first sub-marginal land purchase for retirement and return to grass came in Cimarron County, Oklahoma.[31]

Actually, a similar land purchase program had been in effect since autumn 1934 under the land policy section of the AAA. Federal land purchases for the retirement of crop land were made in southern Otero County, Colorado; Harding County, New Mexico; and Morton County, Kansas. These purchases totalled 153,278 acres; 63,765 acres; and 52,364 acres respectively. In Otero and Harding Counties, a high percentage of the land was already in native grass and was leased back to cattlemen to enable expansion of their operations and to encourage the shift from cash crops to livestock. In southwestern Kansas where the sandhills of Morton County had been badly abused, grazing land was blowing as badly as plowed lands when the purchases were made. Before reseeding grass, the land had to be stabilized with contour plowing and strip cropping. Kansas Senator Arthur Capper supported the federal land purchase program and hinted of the possible need to expand it when, in July 1936 he said, "It may be that thousands . . . of acres should be purchased by the government and retained as public grazing lands, [and] utilized at times for growing crops to meet some emergency that requires abnormal grain production for a year or so."[32]

Congressman Clifford Hope, along with other Dust Bowl residents, favored a similar plan. Some of Hope's constituents urged the federal government to undertake a lease program whereby it would rent the worst blowing land and allow it to remain idle until conditions improved. Others recommended that the government purchase badly eroded lands and lease potentially dangerous wind erosion areas. Although these ideas might be generally accepted by nonresident farmers, Hope wondered what would happen to tenant and resident family farmers. Still, Hope thought, "the government will be willing to go as far as the people in that area will let them go along the line of taking land out of production."[33]

By mid-March 1937, the federal government had accepted options to buy about 53,000 acres of land in Morton County. The land involved was rough, rocky, and sandy. Most of it had never been cultivated. Where purchases were made, the government removed all improvements, even water well casings. The purchases were slow, however, and Hope thought most of the county would blow away before the government would be able to stabilize the land.[34]

In July 1937, the submarginal land purchase program speeded up with the passage of the Jones-Bankhead Farm Tenancy Act. Title III of that act expanded the land purchase program. In part, it authorized the secretary of agriculture to purchase submarginal lands not suited for agriculture in the Dust Bowl. In Colorado, 786,250 acres were to be purchased in Baca and Las Animas Counties. The Kansas purchase area totalled 92,836 acres south of the Cimarron River in Morton County. The government also planned to purchase 161,664 acres in Union County, New Mexico; 288,678 acres in Dallam County, Texas; and 85,452 acres in Cimarron County, Oklahoma. The purchases were to be selective. Only those lands that were beyond the control of private owners were to be taken, and local county planning committees were to designate areas where the purchases could be made. All federal purchases were to coordinate with other agricultural programs designed to help restore the area.[35]

No one knew how much land might be taken eventually, but it could not be taken against the owner's will. If a farmer wished to sell, the government sent appraisers, and both parties tried to arrive at a fair price based on soil conditions and improvements. The farmer retained the mineral rights and the government paid all back taxes. As a large land purchase program began, some Dust Bowl farmers became fearful that the government was intent on moving them out of the area at all costs. The editor of the *Elkhart* (Kansas) *Tri-State News* assured his readers that only "wild land," not under lease or cultivation, which was a wind erosion menace would be purchased. The lands which the government did purchase were to be restored to grass and eventually used for "controlled grazing," wildlife refuges, and public recreation areas. Although that task would require years of work, and although some Dust Bowl residents thought the outcome was "dubious," most agreed that the objective was commendable.[36]

By 1939, however, the submarginal land purchase program had fallen into disrepute in Morton County because the FSA reportedly was buying good lands which forced tenants into the towns and onto the relief rolls. Z. W. Johnson, Morton County agent, wrote, "It appears they [the FSA] are trying to buy the improved places and then just freeze out the others." Local governments also began to fear additional losses to the tax base; others resented the government destroying improvements that were often worth more than the land. Ray Johnson, vice president of the Southwest Agricultural Association, thought these fears were "ridiculous" and a mere "tempest in a tea pot." Actually, the rumors that the FSA was planning to buy all the good land in the Dust Bowl stemmed from the agency taking options on two or three tracts of "fairly good" farm land. The tenants became fearful they would be moved out, and businessmen panicked at the prospect of losing customers.[37]

The real reason for dissatisfaction with the government's land purchase program was, of course, the return of adequate precipitation for crop raising in most areas. In January 1939, Morton County had the best wheat prospect of any county in the state. The war in Europe also forced Congress to curtail many programs in favor of boosting defense. As a result of these two influences, the submarginal land purchase program was terminated in February 1940. By that time nearly 1 million acres of submarginal land had been purchased since 1935 at an average price of $3.56 per acre. Although 250,000 acres were still subject to wind erosion, 350,000 acres had been restored to grass and were under lease to cattlemen.[38]

In retrospect, the Dust Bowl cattlemen's problems stemmed from heat, drought, dust, depression, and overgrazing. Although depression made the cattle situation bad enough, the drought made it nearly hopeless. Indeed, heat and drought alone killed an estimated 50 to 70 percent of the grass on the nongrazed areas. Grazing killed an additional 8 percent, and soil blowing caused more damage. Certainly, the Dust Bowl cattle producers have been unduly criticized for overgrazing their pastures and for helping cause soil erosion. Livestock raising does not permit quick adjustment when climatic or economic changes are swift, adverse, and severe. Dust Bowl cattlemen needed time to locate new pastures in other sections of the country, to ship in feed, and to make marketing adjustments. Herds cannot be disposed of over a few weeks time. If a large number of pregnant cows or calves are on hand, the process may take a year or more. This is not to say that Dust Bowl cattlemen were not guilty of overgrazing and mismanagement of their grasslands, because some were. Still, if the grass does not grow, even understocked pastures are quickly overgrazed. By the time cattlemen had made the necessary adjustments, the rains returned. In some areas, the 1938 feed crop was the best since 1931, but it was expensive to harvest and cattlemen preferred to buy more livestock and graze it off. In Morton County, Kansas, cattle producers and businessmen met to discuss the need to bring more livestock into the area. All of them agreed that additional livestock were "vitally needed" to insure the return of "old time prosperity and a sound farm program."[39]

As they prepared for a rebirth of the cattle industry, Dust Bowl cattle producers, townsmen, and government officials alike began to turn their attention to a panacea. It was an idea which many hoped would break the incessant wind, reduce soil erosion, and end the terrible dust storms for all time. It was nothing less than a plan to bring forestry to the Great Plains.

CHAPTER EIGHT

The Shelterbelt Project

On a hot July day in 1932, Franklin Delano Roosevelt's campaign train was detained near Butte, Montana. As F.D.R. gazed out the window across the bleak, treeless landscape, denuded by copper smelting fumes, it reminded him of the windswept Great Plains. The grim Montana landscape, his own interest in forestry, and the intensifying drought and dust storms in the Plains sparked an idea in his mind. As the train rolled forward, F.D.R. conceived a grand tree-planting scheme for the Great Plains. It was a plan which, he believed, would have two major purposes. First, a giant block of trees stretching from Canada to Texas would ameliorate drought conditions by slowing the force of the wind, reducing evaporation, and protecting soil from erosion. Secondly, it would provide useful employment for the region's jobless. Prior to the election, however, Roosevelt was not prepared to make his idea public. Like so many forthcoming New Deal programs, the plan for the Shelterbelt Project, as it came to be known, was not entirely developed prior to that time. After the presidential election, though, government planning began for a massive tree planting campaign on the Great Plains.[1]

Actually, Roosevelt's idea to plant trees in the Great Plains was not new. Settlers who moved into the Plains during the late nineteenth century often planted trees around their farmsteads. Trees made homesteads more pleasant places to live, provided shade, and gave settlers a sense of permanence. Indeed, the Kansas Horticultural Society noted in its annual report for 1880: "Those settlers who planted shelterbelts and groves are fixtures on their land, while those who never planted trees have pulled up stakes and gone elsewhere." Several years earlier, in 1872, J. Sterling Morton, a Nebraska orchardist, proposed a special tree planting day. Governor Robert W. Furnas supported Morton's idea and, with the support of the board of agriculture, proclaimed the first annual Arbor Day which was held that April. Although the day received little publicity, 3 million trees were planted. The following year, 1873, Willis H. Drummond, commissioner of the General Land Office, recommended that Congress promote tree culture under the Homestead Act by making land claims conditional upon planting a certain number of trees. Drummond's recommendation, plus similar public urgings, encouraged Congress to pass the Timber Culture Act. Under that legislation, settlers could claim 160 acres of public land if they planted 40 acres of trees and cultivated those plantings for ten years. Since most farmers could not tend 40 acres of trees and and still have time to plow, plant, and harvest, Congress amended the act in 1878. The change reduced the required planting to 10 acres with title granted after eight years residence, provided a specific number of trees were still alive. Land entry under the Timber Culture Act was popularly known as a "tree claim."[2]

During the 1870s and 1880s the railroad companies, which were stretching across the Great Plains, also planted trees. The Kansas Pacific; the Atchison, Topeka and Santa Fe; the Burlington and Missouri River; and the Northern Pacific Railroads planted shelterbelts along their lines for snow protection and to make company lands as attractive as possible for potential settlers. Although railroad officials, among others, erroneously believed trees would increase annual precipitation in the Plains, these companies helped to prove that trees could be grown successfully in the grasslands.[3]

Other tree planting efforts followed the creation of Arbor Day, the Timber Culture Act, and the work of the railroads. In 1891, the first seedlings were planted in the Nebraska sand hills. Little more than a decade later, those trees were designated the Nebraska National Forest. In 1905, President Theodore Roosevelt set aside 30,000 acres of western Kansas sand hills south of the Arkansas River for the Garden City Forest Reserve. The Forest Service hoped to duplicate the Nebraska success, and tree planting began. Two years later the Reserve was renamed the Kansas National Forest. But, prairie fire and drought destroyed the plantings, and, in 1915, President Woodrow Wilson abolished it by executive order. The Department of Agriculture had, however, taken a more fruitful measure in 1905 to help increase tree planting in the Great Plains when it created the Office of Dryland Agriculture in the Bureau of Plant Industry. Major dryland experiment stations were established across the Great Plains. Those stations, including the Southern Great Plains Station at Woodward, Oklahoma, conducted forestry research. The dryland experiment stations not only continued the tree planting movement begun in the nineteenth century, but also created a scientific base for tree planting across the Great Plains. In addition, plainsmen gained more tree planting experience after Congress passed the Clarke-McNary Act in 1924. This act authorized the Department of Agriculture to encourage the planting of shelterbelts and woodlots on barren lands through the extension services of the land grant colleges. Although these early tree planting efforts and experiments did not cover the Plains with forests or woodlots, they did set a precedent for future tree planting endeavors.[4]

After the inauguration in March 1933, President Roosevelt asked Robert Y. Stuart, chief of the Forest Service, whether a major tree planting program for the Great Plains would be feasible. As a result of that inquiry, the Forest Service submitted a tree planting program to the president in mid-August 1933. That plan, known as the Shelterbelt Project, called for the creation of a "shelterbelt zone" from the Canadian border to northern Texas. The Forest Service decided to use a shelterbelt zone roughly one hundred miles wide, rather than a solid block planting, in order to avoid too wide diffusion of effort and to obtain the quickest possible results from group plantings. Rexford G. Tugwell, under secretary of agriculture, began coordinating plans for the program's development. By May 1934, the western boundary of the zone had been tentatively set where eighteen inches of precipitation fell annually—a line running approximately from Bismarck, North Dakota, to Amarillo, Texas. Within that zone, shelterbelts 132 feet wide were to be planted on a continuous north-south axis one mile apart across the entire zone. Each belt was to be planted at the center of every section on the "quarter line" fence row which generally separated farm properties. The plan called for planting native trees such as red cedar, hackberry, and green ash at a cost of $34 per acre, excluding land cost. Nurseries in each state transected by the zone would provide the seedlings. The Forest Service estimated the cost of completing the Shelterbelt Project at not less than $60 million over a ten-year period.[5]

Although Roosevelt had not publicly revealed the plans for the Shelterbelt Project, the Associated Press learned of the scheme in June 1934 and began publishing information about it. On June 28, Charles L. Pack, president of the American Tree Association, whom Roosevelt had consulted about the project, revealed the plan, and the Forest Service felt compelled to acknowledge that it was under consideration. President Roosevelt did not sign an executive order allocating the use of $15 million in emergency drought relief funds for beginning the project until July 11, and the project was not officially announced until July 21. The estimated cost had by that time been revised to $75 million over a twelve-year period. The Forest Service estimated that $2.5 million from the initial allocation would be spent during the first twelve to eighteen months. That expenditure would pay farmers for soil preparation, fencing, and cultivation of the trees. The

Forest Service established the Shelterbelt Project's field headquarters in Lincoln, Nebraska, and state offices were also created. In the Dust Bowl states those offices were located at Manhattan, Kansas; Oklahoma City, Oklahoma; and Wichita Falls, Texas.[6]

No sooner had the Forest Service announced the Shelterbelt Project than the foresters took sides over the plan. Some critics contended that trees could not be grown on a large scale in the Great Plains. Others claimed that the project was too costly and that the money should be spent for more "assured" forestry projects in other sections of the country. Some foresters clearly feared that the project would fail and discredit their profession. One forestry expert revealed that apprehension when he wrote: "For some thirty years the lumbermen have thought of the professional foresters as a lot of rattle brained theorists, and we are just now beginning to disabuse them of the opinion and secure their confidence. And now are we going to lose the ground we have thus gained by sponsoring, or perhaps I should say condoning, a project of the magnitude that most of us feel is doomed to failure?" Another wrote, "We might conceivably cover the High Plains with trees, and we might carpet the state of Maine with buffalo grass, but if we are sensible we shall try to do neither." For these opponents and others like them the Shelterbelt Project would probably boomerang and give forestry a "terrific setback in public opinion."[7]

Foresters who supported the project pointed to previous tree-growing successes since passage of the Timber Culture Act, and they particularly noted the work of the dryland experiment stations. At the Northern Great Plains Field Station near Mandan, North Dakota, for example, the Bureau of Plant Industry had been working with farmers since 1916. As a result of that work, more than 2,700 demonstration shelterbelts had been planted by 1933, and the survival rate averaged 70 percent. H.H. Finnell, director of the Dalhart Wind Erosion Control Project, supported the Shelterbelt Project. He believed that if a consistent tree planting program had been conducted prior to the 1930s, much of the soil erosion in the Texas Panhandle would have been prevented. Other supporters claimed that the shelterbelts would check evapo-ration; protect crops, livestock, and farmsteads; and increase property values.[8]

Generally, Dust Bowl farmers were not at first enthusiastic about the project. Some believed only the next generation would benefit from it, while current farmers needed protec-tion from wind erosion immediately. These farmers did not believe the shelterbelts would end the dust storms, although some thought the treebelts would protect local field crops. Most Great Plains editors favored the project, even though the cost seemed astoundingly high for this fiscally conserva-tive region. The *Amarillo Globe* noted, however, that if drastic steps were not taken to end the wind erosion menace, the Great Plains could become another semiarid wasteland like China.[9]

Financial problems plagued the project from the beginning and prevented the Forest Service from conducting the plant-ing operation on the scale it had planned. The project received an almost immediate setback when John R. McCarl, comptroller-general, a Harding appointee, refused to release the project's funds. McCarl took the position that the plan was not an emergency relief measure but was rather a long-range recovery program. Since Congress had provided the money for immediate relief, McCarl contended that congres-sional approval was needed for this program. Therefore, the project was not entitled to the emergency drought relief funds. He did consent, though, to release $1 million to allow the Forest Service to begin work, and the planning continued.[10]

Indeed, planning work intensified. In order to satisfactorily answer the critics, the Lakes States Experiment Station was assigned the task of determining whether shelterbelts could grow in the Great Plains. As a result of that mandate, the experiment station conducted studies during the latter half of 1934 and early 1935 on climate, soils, native vegetation, and earlier tree plantings in the six-state shelterbelt area. The report that followed, written by Raphael Zon, director of the Lakes States Experiment Station, established the basis for the 1935 planting season. Zon reaffirmed that tree planting in the Plains was possible, identified the species for planting, and refined the project's boundaries.

Zon recommended that the zone be delineated on the west by a line which marked the area receiving sixteen inches of annual precipitation and the eastern boundary be restricted to the twenty-two inch annual rainfall line. The western boundary roughly followed the ninety-ninth meridian from Devils Lake, North Dakota, to Mangum, Oklahoma, and created a zone including eastern North and South Dakota, east-central Nebraska, west-central Kansas, western Oklahoma, and a portion of the Texas Panhandle. In reality, the boundaries merely delineated the area within which tree planting would be most successful according to soil types, annual precipitation, land use, and need. Not all of the soil types found in the zone would support tree growing. Indeed, only 56 percent of the 114,700 square mile area would lend itself favorably to shelterbelt planting. Trees could be grown on another 40 percent of the soils only with difficulty, while 4 percent of the area was entirely unsuitable for trees. Therefore, continuous 1,000-mile-long shelterbelts spaced one mile apart would be impractical. Instead, each planting would have to be adapted to the soil types found on individual farms. Furthermore, Zon also advised the Forest Service to plant the shelterbelts in an east-west direction which would be at right angles to the prevailing winds, rather than according to the original north-south plan.[11]

Zon contended that "highly stabilized control of the land dedicated to the shelterbelts is essential." To insure that the shelterbelts would be properly cared for, Zon recommended maintaining control in four possible ways—government ownerships through purchase or donation, grants of perpetual easements, leases, and cooperative agreements between the federal government and the land owners. For the 1935

planting season, the Forest Service decided to lease the lands
for the shelterbelts, however, the government reserved the
right to purchase land at any time during the life of the lease.
The details for the land lease plan were borrowed, in part,
from the Agricultural Adjustment Administration's allotment
program in which farmers were compensated for retiring land
from production. Forest Service representatives negotiated
the lease agreement with the landowner and the annual rental
fee was based on the productive capacity of the land. During
1935—the only year this procedure was used—rental payments averaged $3.67 per acre. After the lease was signed for
a strip of land not less than ten rods wide, the farmer was
usually hired to plow the land for planting, fence the area, and
cultivate the trees. The actual planting was conducted by Forest Service crews.[12]

The Forest Service planned to get 4 million seedlings for
the 1935 planting season from commercial and government-
owned nurseries. Half of that planting stock was scheduled
for shelterbelts in a twenty-four county area in the Dust Bowl
states. Considering the overall scope of the project, this
planting was to be small and largely demonstrational. The primary task was to plant the trees and keep them alive for a
year, during which time the Forest Service hoped to gain
additional support from the farm community and Congress.
The planting was conducted at a fervid pace. More than ten
thousand trees were planted in Comanche County, Kansas, in
a week. By the end of April, fourteen miles of shelterbelts
were planted in Oklahoma and nearly twenty-five miles
established in Kansas. Those belts averaged between a half
mile and a mile long and were 132 feet wide. Almost a dozen
species were planted in each belt with the tallest-growing
trees located in the center and the smaller shrubs along the
outside to help lift the wind. The Texas plantings, however,
did not fare as well. Only two one-half mile strips were
planted in Childress and Wheeler Counties. There, organiza-
tional difficulties and farmer reluctance to participate
retarded the program.[13]

The first planting season was difficult at best for the men working in the fields. The spring of 1935 was the worst Dust Bowl year. Extensive drought, intensive heat, nearly constant wind, and black, rolling dust storms worked against man and the seedlings. Occasionally the dust was blowing so badly that planting operations ceased for several days on end. Consequently, from the planting of the first tree on March 19, near Mangum, Oklahoma, until the planting season ended on July 1, only 120 miles of shelterbelts had been established for the entire zone. But, 4,800 acres of wind breaks had also been planted around 1,800 farms for a total cost of less than $697,000. Nevertheless, a start had been made, and the Forest Service immediately began preparations for the 1936 season.[14]

The Forest Service hoped to plant about 56 million seedlings during spring 1936. In order to have that number of trees ready, the Forest Service leased 552 acres, mainly from private nurserymen. These government-operated nurseries were necessary because commercial nurserymen were reluctant to provide planting stock since the Forest Service could not pay in advance or contract for deliveries after the end of a fiscal year. Commercial nurserymen did not want to risk financial losses if they had to depend upon yearly congressional appropriations to reimburse them for trees already grown and investments made a year earlier. Also, government nurseries provided relief work. The Forest Service established fourteen nurseries in the spring of 1935; at least one nursery was located in each Great Plains state within the zone.[15]

Prior to the 1936 planting season, the Forest Service also modified its policy for leasing lands because of public criticism that farmers received duplicate rental payments from the Shelterbelt and AAA programs. As a result, these two agencies agreed that the Agricultural Adjustment Administration would continue to pay farmers for retiring land from production and that the Forest Service would, with the farmer's consent, plant shelterbelts on it. Only farmers who participated in the AAA program could have shelterbelts under the Forest Service's program.[16]

This plan was never activated, because in January 1936 the Supreme Court held the AAA unconstitutional. Another cooperative agreement was, therefore, hastily devised. According to that plan, the property owner agreed to furnish the land, prepare the soil for planting, provide fencing materials, cultivate the seedlings, and share the cost of pest control. For its part, the Forest Service agreed to supply the planting stock, plant the trees, and fence the shelterbelts. The landowner was also allowed an additional windbreak, not to exceed three acres, to be maintained at his expense. As a result of this agreement the landowner and the federal government divided the expenses on a nearly equal basis.[17]

Although soil moisture conditions had been less than satisfactory during 1935, the survival rate of the trees planted in the Dust Bowl states averaged between 60 and 70 percent. This high rate of success, plus an additional allotment of $1,990,958 from the Emergency Relief Appropriation Act of 1935, encouraged the Forest Service to plan approximately 1,400 miles of shelterbelt and 6,400 acres of farmstead windbreaks for 1936. But, since the Works Progress Administration (WPA) administered the Emergency Relief Appropriations Act, the agency imposed several restrictions upon the Forest Service. First, the WPA required that 90 percent of the funds be used to hire relief workers; not more than 10 percent of the allocation could be used for equipment or administrative expenses. Second, these funds were to be divided among the states, which prevented shifting funds from one state to another as requirements dictated.[18]

To better organize the upcoming planting season under these new restrictions, the Forest Service divided the states into districts. Kansas, Oklahoma, and Texas had six, four, and three districts respectively. The staff of each district consisted of a shelterbelt assistant, a foreman, and a planting crew. When the 1936 planting season ended, 2,212 shelterbelts and 878 farmstead windbreaks had been established in the six-state zone. A dramatic increase in plantings had been made in the Dust Bowl states. In Texas nearly 1,767,000 trees stretched 171 miles in six counties. An additional 200 acres of "farmstead strips" protected 150 farms with more than 100,000 trees. The Forest Service achieved similar results in Oklahoma and Kansas. In the Sooner State, 157 miles of shelterbelts including 1,281,694 trees were planted. The Kansas results were even more impressive. There planting crews established over 215 miles of shelterbelts containing 3,287,700 trees; an additional 178,925 trees protected 146 farmsteads. Shelterbelts now totalled 1,152 miles on 1,936 farms in the Dust Bowl states. By July 1, 1936, the survival rate ranged from 76 percent in Texas to nearly 80 percent in Oklahoma. Those results convinced the Forest Service that it had demonstrated beyond question the "entire feasibility" of the Shelterbelt Project.[19]

Congress did not agree with that assessment. In June, Congress refused to appropriate additional funds for the project and allocated $170,000 for its liquidation. Roosevelt was unwilling to allow one of his favorite programs to be terminated and instead authorized the Shelterbelt Project to continue under WPA funding. With the WPA sponsoring the 1937 planting season, several changes were made in the program. First, the planting sites were no longer restricted to the original zone. Shelterbelts could now be planted as far west as was feasible to help protect the areas where the need was greatest. Secondly, farmers were required to sign a contract with the Forest Service in which they promised to donate the land, build and maintain fences around the plantings, cultivate the seedlings, and help control rodents. The Forest Service in turn agreed to supply the stock, plant the seedlings, and help to cultivate and build fences whenever funds were available. The Forest Service reserved the right to inspect the trees at anytime, and farmers were prohibited from removing the trees. If a farmer breached the contract, the Forest Service could terminate the agreement. Under this policy, the landowner paid about 60 percent of the cost for establishing a shelterbelt. The Shelterbelt Project was now officially rechristened the Prairie States Forestry Project.[20]

With new agreements signed and funds available for another year's work, tree planting resumed in the spring of 1937. In Oklahoma, 1,849,000 trees were used to plant nearly 325 miles of belts. The Texas plantings totalled more than 1,340,000 trees in 225 miles of treebelts. In Kansas, nearly 1,341,000 trees were used in more than 202 miles of shelterbelts. In the zone as a whole, 1,321 miles of shelterbelts were planted at a cost of $1,730,650.[21]

Still, a long-range program could not be planned with only tenuous WPA funding. Therefore, Secretary Wallace again sought congressional recognition and annual funding for the project. To accomplish that goal, Wallace persuaded Roosevelt to seek a comprehensive farm forestry act, which would allow the Forest Service to continue the shelterbelt program. The president supported the idea and, with his influence, convinced Congress in May 1937 to pass the Norris-Doxey Act, which is commonly known as the Cooperative Farm Forestry Act. This act gave the Prairie States Forestry Project "functional authorization," but little else. Specifically, Congress authorized the secretary of agriculture to cooperate with the land grant colleges and state foresters for providing planting stock, for entering into cooperative agreements, and for establishing nurseries. Although the Act authorized $2.5 million in annual appropriations, no money was forthcoming. The House Subcommittee on Appropriations refused to commit funds to a project that would not show significant tangible results for many years to come. As a result, the WPA financed the 1938 planting season.[22]

Although congressional support remained almost nonexistent and WPA funds were limited, grass-roots support for the project began increasing. Of the three Dust Bowl states, Kansas showed the greatest enthusiasm. Farmers, county commissioners, farm organizations, newspaper editors, women's organizations, and educational leaders gave increasing attention to the project. Because of that support, T. Russell Reitz, state director of the Prairie States Forestry Project, wrote, "The place of trees in Kansas agriculture is slowly but surely gaining recognition." Indeed, so many Kansas farmers were requesting shelterbelts that the Forest Service began organizing township tree-planting committees during the summer of 1937.

The Forest Service selected committee members, with the advice of the county agent, from the leading farmers interested in soil conservation. The committees were given the responsibility for locating the shelterbelts in a manner that would be most beneficial to their townships. Possible shelterbelt locations were discussed at community meetings, and landowners were urged to cooperate. The township tree committees stressed, as did the Soil Conservation Service, that success depended upon everyone assuming a fair share of the responsibility to halt wind erosion. The Forest Service believed that this committee plan would encourage farmers who were skeptical of federal bureaucrats to participate in the project. The first committees were established in Stafford and Edwards Counties in September; by the end of 1940, 240 township committees had been organized.[23]

By the 1938 planting season, the Forest Service was greatly impressed with the results of the work completed over the previous three years. In Kansas, trees planted in 1935 were already eighteen to twenty feet tall and were already protecting crop land from wind erosion; trees planted a year later were twelve to sixteen feet high. In Texas, 80 percent of the 13,000 trees planted in the Dalhart area since 1935 had survived. When the spring planting season ended, nearly 7.5 million trees in 1,044 miles of shelterbelts protected Oklahoma soil. In Texas more than 5 million trees formed 768 shelterbelt miles, and in Kansas 4 million trees composed nearly 700 miles of tree belts. As those trees grew, they shielded the wheat fields from the wind and retarded soil erosion, and additional belts were established in eastern Colorado and northeastern New Mexico—far beyond the original shelterbelt zone. By midsummer 1938, absentee landlords from

Pennsylvania, Missouri, Illinois, and California were applying to the Forest Service for shelterbelt agreements. With money in short supply, however, the Forest Service now required landowners to perform all cultivating of the season's plantings in addition to supplying the land free of charge, preparing it for planting, and providing fencing material. Nevertheless, farmer cooperation remained high. In several Kansas counties, all of the participating farmers provided the required cultivation, and no county registered less than 98 percent cooperation in that endeavor.[24]

Although grass-roots support was increasing, Congress was still reluctant to fund the project. Nursery owners opposed the Prairie States Forestry Project during the 1938 hearings on the grounds that government nurseries were undermining their businesses. That argument was unwarranted because, prior to the initiation of the Shelterbelt Project, Great Plains farmers purchased few trees from either commercial nurseries or state extension services. Actually, instead of losing business because of the federal project, commercial nurserymen improved their businesses because Plainsmen now became more tree conscious. Commercial nurserymen did not really oppose tree planting in the Plains; rather they wanted their share of the government's business and were eager to supply planting stock. Therefore, they requested Congress to pay farmers for planting trees which could only be purchased from commercial nurseries. The Forest Service favored purchasing half of the planting stock from commercial nurseries, but, since Congress continually failed to provide dependable annual appropriations, it could not do so. In addition to the opposition of the nurserymen, Congressman Wall Doxey, one of the sponsors of the Farm Forestry Act, gave a fatal blow to Forest Service hopes for congressional funding. At the appropriations hearings, Doxey testified that he had not intended that legislation to benefit the Prairie States Forestry Project. As a result, Congress once again denied funding, and the WPA continued to finance the project for the remainder of its life.[25]

In 1940, Harold Smith, director of the budget, suggested shifting the Shelterbelt Project to the Soil Conservation Service. Since Congress had already authorized the SCS to use its funds for tree planting activities, Smith reasoned that the project would have a better chance of annual appropriations if it were linked to that agency. Furthermore, the Plains states had soil conservation districts organized which could provide the necessary planning. Roosevelt accepted that idea, and, in early May, the Prairie States Forestry Project began its transfer to the Soil Conservation Service. The continual lack of funding was not the only reason for the removal of the project from WPA sponsorship. The war in Europe had stimulated the economy, and fewer men needed employment on federal projects. Also, with the drought broken, the Forest Service could no longer justify the project's continued existence. Consequently, on July 1, 1942, the project was officially transferred to the SCS. Shelterbelts then became part of the Soil Conservation Service's program to use all appropriate measures to halt soil erosion in the Great Plains. The Forest Service then relegated itself to supplying planting stock through the state forestry services and extension agencies.[26]

Although the Shelterbelt Project was terminated at that time, the Forest Service had achieved a remarkable success in tree planting, particularly in the Dust Bowl states. Overall, during the project's eight-year history, the Forest Service planted nearly 18,600 miles of shelterbelts with 217,378,352 trees. A total of 30,223 farmers participated in the program at a total cost of $14,862,307. In the Dust Bowl states of Kansas, Oklahoma, and Texas, shelterbelt mileage totalled 3,540; 2,994; and 2,042 miles respectively. In 1944, the survival rate for trees planted between 1935 and 1938 in those three states ranged from nearly 60 percent in Oklahoma to 74 percent in Kansas. For the entire zone, the survival rate averaged 58 percent. Of course, the shelterbelts grew best in areas where the soil was sandy and where there was a high percentage of owner-operated farms. Farm owners generally took

better care of the seedlings by cultivating at the proper time
and by keeping cattle out. Tenant farmers tended to be negligent in their upkeep of the treebelts.[27]

The outbreak of World War II and the transfer of the Shelterbelt Project to the SCS caused an irreparable setback for
the program. The war caused a deficiency in manpower, gasoline, and replacement equipment which prevented farmers
from giving the treebelts adequate cultivation. With the
return of normal precipitation during the early 1940s, many
farmers lost interest in continued maintenance and planting.
If the dust was not blowing and crop production was high,
they could see no reason to do so. Upkeep deteriorated, and,
in the Dust Bowl states, cattle damages were greater than in
the other Great Plains states. Oklahoma and Texas farmers
became particularly negligent in shelterbelt maintenance.
The SCS must bear some of the responsibility for the deterio-
ration of the project's work, because that agency did not
emphasize tree planting except for farmstead and livestock
protection. The SCS preferred to use other methods to control wind erosion. Although the SCS continued to supply
technical assistance for species selection, soil preparation, and
planting techniques, it had only a limited supply of planting
stock. The SCS expected farmers to purchase their seedlings
from state nurseries. Under these conditions, shelterbelt
plantings virtually ceased. Still, by the mid-1950s, 73 percent
of the trees were still alive and were effectively checking soil
erosion, protecting farmsteads and livestock, and providing
wildlife shelters.[28]

During the 1950s, Dust Bowl farmers, particularly in Texas
and Oklahoma, began removing the shelterbelts. These farmers contended that they could control wind erosion better
with modern equipment than with shelterbelts. They also
argued that the land taken by trees could be more profitably
used for crop land and that the trees took valuable moisture
from adjacent fields. These attitudes were especially preva-
lent among younger farmers who had not lived through the
dust storms. Tenants also disliked the shelterbelts because

they resented having to divert acreage from production even though the trees helped to reduce soil erosion on their crop land. That acreage loss was negligible, however, since most farms had only one shelterbelt, consisting of ten rows of trees which extended from one-eighth mile to a mile in length. The older farmers often felt Dust Bowl conditions might return and preferred to keep their shelterbelts. Some farmers also argued that the shelterbelts did not protect more than a small area immediately adjacent to the trees. Indeed, shelterbelts could not reduce wind velocity for distances beyond twenty times the height of the trees. Still, many farmers claimed and experiments proved that shelterbelts helped to reduce soil blowing, slowed evaporation, and increased crop yields.[29]

Certainly, the Shelterbelt Project was never the panacea for ending the dust storms that many had hoped it would become. Forest Service experts realized from the inception of the project that the treebelts would not end the dust storms. Rather, the primary purpose of the Prairie States Forestry Project was to grow trees in order to reduce wind erosion, stabilize the land, and make the Plains "a better and more profitable place to live." The Forest Service did not intend to withdraw large blocks of land from agricultural production or to transform the Plains into a forest. Instead, the project was designed to help increase agricultural productivity of the protected lands. In this respect, then, the shelterbelts were only part of a larger conservation program. Still, when the Shelterbelt Project was combined with the work of the other government agencies which fostered land retirement, controlled grazing, farm pond construction, strip cropping, terracing, and agricultural diversification, it made a major contribution to the physical and psychological fight against the wind erosion menace. During the early 1950s, however, just as many Dust Bowl farmers were beginning to destroy their shelterbelts, the drought returned and with it came the dust storms.[30]

CHAPTER NINE

*Return of the Dust Bowl**

Throughout the 1940s, farmers and ranchers in the southern Plains prospered from bumper crops and the high prices for wheat, cotton, and beef brought on by wartime demands. New farmers, wanting to capitalize on such favorable conditions, moved into the old Dust Bowl and began to work much of the land abandoned during the 1930s. The lush grass and bountiful crops made the black blizzards of the previous decade seem like ancient history. Good rains appeared normal and drought was all but forgotten. One observer warned: "When it turns dry again, the dust will blow again. The intensity of the black blizzards will depend on how long the drought lasts, over how large an area, no matter whether man leaves or stays, whether the land is plowed or grassed, whether there are more or less trees." Few people paid any attention and land values soared. Farms that could not be sold for $3 to $4 per acre in the 1930s now brought $40 to $60 per acre. As a result, during the ten-year period from 1941 to 1950, another big plow-up occurred on the edge of the Dust Bowl, where annual precipitation averaged fourteen to seventeen inches. There, the sod was broken for wheat in Colorado

---

* Reprinted with permission from the October 1979 *Journal of the West.* Copyright 1979 by the Journal of the West, Inc.

and for cotton in west-central Texas and eastern New Mexico. Farmers broke about 4 million acres at that time, 3 million of which were submarginal and unfit for cultivation. In 1945, Cheyenne County, Colorado, had 229 fewer farms than ten years earlier, at that time, though, it had 931,142 acres under cultivation compared to 512,424 in 1935. One farm editor warned that a potential Dust Bowl was in the making, but he too went virtually unheard.[1]

A few dust storms occurred in the southern Plains during the mid-1940s but most were local in nature and only scattered fields were damaged. The sandy land south and west of Lubbock, Texas, was an exception. By 1947, sand storms were common on those newly broken cotton lands. Some of the sand storms were severe, and wind erosion scoured many fields. At the same time, the rains became less frequent. By late December 1948, western Kansas and eastern Colorado wheat lands as far south as the Oklahoma Panhandle were blowing. Not since the 1930s had the soil moved so badly. In Greeley County, Kansas, where most of the land was owned by nonresidents, the county commissioners invoked the soil drifting law. Notices were sent to several farmers ordering them to work their blowing fields or the county would do it for them. Farther to the west, a rancher in Kiowa County, Colorado, reported that the soil was drifting so severely across the tracks of the Missouri Pacific Railroad that the trains had to stop while the crews shoveled the tracks clear. Indeed, so much land had been plowed in eastern Colorado that not enough grass remained to anchor the soil in case a prolonged drought returned. At that time, one western Kansan prophetically warned, "We aren't kidding ourselves that it won't come again. There isn't a thing being done to prepare for it. If it starts again, within two or three years . . . it will catch us as ill prepared or almost so as it did in the middle thirties. Only this time it will be much more widespread with so much more land broken up out there in Colorado." Colorado farmers were not alone in their negligence. The SCS reported that 2.4 million acres were "plowed out" in west

Texas and western Oklahoma. In short, scant vegetation remained to anchor the soil.[2]

By mid-January 1950, soil blowing was becoming worse in the southern Plains. The top soil was by now "powder dry," and strong winds moved it easily. In Colorado and the Oklahoma Panhandle, several areas were blowing badly. With the coming of spring, the approach of another drought was apparent to most Dust Bowl residents. The wheat crop was already suffering from lack of moisture, and many fields had blown out where cattle had been allowed to overgraze during the winter. In early March, fifty- to seventy-mile-per-hour winds brought a return of the dreaded dust storms which, according to a local editor, "were hard on everyone's disposition." Most people thought that the rains surely would come soon, but the rains did not come. Southwestern Kansas received less precipitation during the first three months of 1950 than during any other quarter year on record, and now the dust storms became more frequent, intense, and widespread than at any time since 1938. Once again, dust storms like those of the 1930s began to sweep beyond the region of the old Dust Bowl. On April 10, while blowing dust reduced visibility to less than two miles at Dodge City, there was zero visibility at Salina in central Kansas. By mid-April, the winds increasingly dried the soil, and Dust Bowl conditions were reported in all but the southeast and south-central portions of Oklahoma. Wheat losses at that time were estimated at $275 million. As a result of the steadily worsening agricultural conditions, one Oklahoman noted, "If this keeps up and the dust comes again, generally, I'll be a C.I.O. You know, a California Improved Okie."[3]

By summer 1950, drought had clearly returned to southwestern Texas and southeastern New Mexico. The following year, the drought extended across most of the southern Plains; by 1952 it was severe in the Dust Bowl. Soil blowing now became a menace on the newly plowed lands, on the poor wheat lands where no crops had been raised for three or four years, and on the poorly managed grazing lands. In early

January 1952, blowing dust and zero visibility caused a six-car crash on the highway between Springfield and Lamar, Colorado. The following month eighty-mile-per-hour winds lifted dust to an altitude of twelve thousand feet in western Kansas, and visibility dropped to two blocks in Oklahoma City. On one occasion, the dust became so thick that it sifted into the Kansas Wesleyan University basketball gymnasium in Salina, during a game with McPherson College. The game was played in spite of thick haze, although the players had to wipe their feet frequently to keep from sliding on the dust-covered floor.[4]

By late February, the agricultural situation was critical in eastern Colorado. A full 70 percent of the 500,000 acres of wheat planted in Baca County had blown out. All grass and crop lands needed immediate listing. Oakley Wade, state representative from Las Animas County, reported that 70 percent of the wheat crop had been destroyed in neighboring Bent County. In Kiowa County 75 percent of the 300,000 acres seeded in wheat were gone; 500,000 acres of cultivated land and 50 percent of the grassland needed to be worked. In Prowers County, 250,000 acres, or 95 percent of the wheat, had blown out or silted over. Fences had blown down or were drifted over in Crowley and Lincoln Counties. Nearly 1.5 million acres were badly damaged in Colfax, Lea, Harding, Roosevelt, Torrance, Union, Quay, and Curry Counties in New Mexico.[5]

In early March, wheat losses were estimated at more than 7 million bushels and the monetary loss at over $14 million in a seven-county area in the heart of the Colorado Dust Bowl. Fred Meyers, a Hamilton County, Kansas, conservationist, estimated his county's wheat loss at 95 percent or about $1.5 million. In addition to suffering these crop losses, soil blowing filled many fields with shifting dunes three feet high. The land was badly hummocked in some areas. One farm couple returned home from a vacation to find an inch of dust covering the floors of their home. A reporter wrote: "The actual physical discomfort is not so bad because they are getting used to it. . . ." A western Kansas farmer remarked, "We'd probably give this land back to the Indians . . . if we had any Indians."[6]

The drought continued. By May 1954 the Dust Bowl had received, on the average, less than one inch of precipitation since the first of the year. Some localities had no moisture at all. Unusually warm temperatures and high winds accompanied the drought. On February 19, the most severe wind storm of the decade struck this region. At Dodge City, forty-nine-mile-per-hour winds blew for six consecutive hours; gusts reached between eighty and ninety miles per hour. After the storm at Garden City, Kansas, the dust had to be shoveled from sidewalks; snow followed the dirt storm and turned the dust to mud. High winds continued throughout the region for the remainder of the spring. By June 30, the Garden City agricultural experiment station had recorded thirty dust storms since the duster on February 19. Several inches of top soil had been removed from nearly all the unprotected fields from southwestern Nebraska and northeastern Colorado to western Texas and southeastern New Mexico. By April 1, wind erosion had damaged about 11.7 million acres of crop land and approximately 5.2 million acres of range land in the southern Plains. An additional 8.2 million acres of crop land and nearly 6.7 million acres of range land would be subject to wind erosion if rain did not come soon. Some farmers estimated that as much as 30 percent of their grain had blown out of the heads before harvest.[7]

Severe blowing continued into the spring of 1955, reducing visibility to zero at times, drifting soil along fence rows, piling sand dunes twenty to thirty feet high in some fields, ruining crops, and scouring paint off license plates until they shone like "polished steel." During one March storm, air conditioners were turned on in Oklahoma City businesses to filter the air. The weather bureau called another dust storm a "real toughie," when dust extended from Denver to San Angelo, Texas, and from the Guadalupe Mountains near El Paso to Wichita. Over a million acres of wheat had been damaged by wind erosion in Kansas and Colorado by March 1; hay and sorghum crops failed during the summer, and by the end of the year over 42,000 acres of wheat in western Kansas and eastern Colorado had been seriously damaged by the wind. Dust Bowl farmers once again took jobs in nearby towns, and journalists branded the southern Plains the Dust Bowl.[8]

Drifting dust continued to be a problem into the spring of 1956. In late March, sixty-mile-per-hour winds blew down telephone lines, billowed dust across 40,000 square miles, and ruined 75 percent, or 3.4 million acres, of Colorado's wheat crop. This storm closed Highway 50 east of Syracuse, Kansas, after blowing dust caused a four-car crash that injured seven people. When the accident victims were taken to the Lakin, Kansas, hospital, their treatment was delayed when dust sifted into the X-ray machine causing it to malfunction.[9]

On April 2, a major duster swept over portions of New Mexico, Colorado, Oklahoma, and western Texas. The editor of the *Big Spring* (Texas) *Herald* called it the "grandfather of dust storms" which made the day "dark as night." Zero visibility was reported at Midland, Big Spring, Childress, Pecos, and Abilene, Texas. In Lubbock, the duster smashed plate glass windows. Periodic dust storms, though not quite as bad as this one, continued into the autumn. In late October, a duster blew much of the newly planted wheat out of the ground in central and western Kansas. At Abilene, Kansas, the dust was so thick the dogs could not see the rabbit at the National Fall Coursing Meet and the races were postponed. By spring 1957, though, the rains began to return, and the drought was broken. Still, occasional dust storms continued periodically to annoy southern plainsmen.[10]

Early in this drought period, twenty-six Prowers County farmers, needing financial help, applied to the Production Marketing Administration for emergency listing aid. The request was answered quickly with a $124,445 allotment for 1951 and an additional $98,645 for 1952. Baca County also received similar funds for those years totalling $147,906 and $182,700 respectively. In Morton County, Kansas, the Agricultural Stabilization and Conservation Service office approved a deep plowing program for farmers with the cost to be divided between the farmer and the federal government. With this help, farmers set their one-way disks fourteen to twenty-four inches deep and tilled about 5,000 acres of sandy land. Together with listers and chisel plows practically all of the idle land subject to wind erosion was treated and the blowing hazard greatly reduced.[11]

Dust Bowl farmers were quick to begin an emergency tillage program. Even if some of them had been negligent over the past decade, and even though the suitcase farmer contributed to the problem, resident farmers soon recognized the seriousness of the new wind erosion menace and began to implement soil conservation procedures they had learned in the 1930s.

In March 1954, the Colorado legislature passed a stringent wind erosion law which set up a $1 million emergency fund for Colorado's Dust Bowl farmers. On the national level, the U.S. Department of Agriculture allotted $2.5 million to the Dust Bowl states for emergency tillage. An additional $12.5 million was also quickly forthcoming. These funds, however, could not be used to reimburse farmers who had begun cleaning out fences, working down hummocks, and applying emergency tillage prior to April 1, when this money became available. Since USDA officials did not have the distribution, regulating, or monitoring procedures worked out for the program prior to that time, they opposed retroactive payments. Unfortunately, many Dust Bowl farmers had completed most of their emergency tillage work before April 1 so did not get government aid to help defray the cost. Even though protests abounded, the Department of Agriculture did not recant. However, most Dust Bowl farmers did not need financial aid. Had the nation been gripped with a depression similar to that of the 1930s, the federal government would have found the farm community less financially secure and probably would have had to initiate an aid program far larger than it did. Nevertheless, by not paying farmers retroactively in 1954, the Department of Agriculture's policy penalized good farmers who practiced the appropriate soil conservation techniques.[12]

Even though Dust Bowl farmers applied emergency tillage to nearly 5.3 million acres by April 1, 1954 (largely on their own initiative), the $15 million in federal aid permitted additional conservation work to be done. Under that allotment program, each farmer who applied could receive a maximum of $1,500—seventy-five cents an acre for listing, fifty cents an acre for chiseling, and $1.25 an acre for contour strip cropping. Payments were also made for the purchase of feed and hay. Furthermore, the Farmers' Home Administration granted emergency loans, and, by August 1954, the FHA had taken over most of the drought relief programs. The federal government spent only $7.9 million of the original $15 million during 1954. In 1955, however, after 11 million acres were damaged by wind erosion between September 1954 and April 1955, the federal government released an additional $5 million to help farmers to continue the battle against wind erosion.[13]

By March 1, 1955, wind erosion had damaged about 4.7 million acres, two-thirds of which was located in eastern Colorado. Much of that acreage was poorly protected from the wind because of inadequate vegetative cover, lack of crop residue, and insufficient moisture. An additional 17 million acres were in a condition to blow. A month later, the wind erosion hazard was remarkably worse. Schools closed in the Oklahoma Panhandle when the dust became so bad that the buses could not run. In southeastern Colorado, the weather bureau reported "blowing stones" during one dust storm. As the drought and dust storms continued, emergency tillage continued too. By April 1, 1955, Texas farmers had treated 3.5 million acres, and Kansas and Oklahoma farmers had applied emergency tillage to 1.5 million acres in each state.[14]

With the wind erosion condition growing progressively worse, the USDA released an additional $4,275,000 for emergency tillage to supplement the $5 million which President Eisenhower had already made available. Of that money, twenty-two Colorado counties received $1,250,000; thirty-eight Kansas counties obtained $900,000; fourteen New Mexico counties were allotted $200,000; five Oklahoma counties received $350,000; and seventy-four Texas counties were allocated $1.5 million. The remaining funds were distributed to Wyoming farmers. With that money, the emergency tillage program continued.

The wind erosion situation persisted. Precipitation for April was below half the normal average. The soil remained dry, and dust storms occurred throughout the month—some lasting thirty-six hours with gusts up to seventy-five miles per hour. As a result, by May 1955, more than 12.9 million acres had been damaged in the Dust Bowl states, and an additional 13.5 million acres were in a condition to blow. Approximately 2.7 million acres of wheat had been injured in Kansas and Colorado alone. The final SCS report for the 1954–1955 blow season shows more than 13 million acres damaged from wind erosion—nearly 6.5 million acres were in Colorado, two-thirds of which was cropland. However, late May rains provided sufficient moisture for sorghum and other soil-holding crops, and, by the end of the year, the wind erosion problem had improved. Still, over 16 million acres were in a condition to blow.[15]

While farmers and ranchers were applying emergency tillage to their lands, they also gave renewed attention to reseeding pasture and range lands. Reseeding was not only a good conservation measure, but also a good business practice, since grass is the only crop that can be grown safely in about two-thirds of the southern Plains. Furthermore, during times of high livestock prices, cattlemen often receive higher returns from grass than from wheat, sorghum, or cotton, and the best grasses for livestock grazing are also the most suitable for soil conservation.[16]

By 1955, much of the southern Plains grassland was in poor condition. Approximately 6 million acres of crop land and 1.8

million acres of range land needed reseeding. In addition, 24,000 miles of waterways and 33,500 dam and farm pond spillways needed a protective grass covering. To meet those needs, the USDA recommended pitting to help establish grass on barren range and pasture lands. Pitting was essentially identical with the basin listing of the 1930s. In order to pit the land, farmers set the disks of their one-way plows off center. As the disks rotated, pits three to four feet long and several inches deep were gouged into the soil. These depressions caught rain water and gave the grass seed a good bed for germination. In addition to blue stem, gramma, and buffalo grasses, the SCS also recommended a mix of crested wheat grass for improvement of submarginal and deteriorated grasslands. This species gave a high forage yield from spring to fall, could withstand heavy grazing, and enabled good weight gains for livestock. Crested wheat grass was particularly well suited for soil conservation in the Dust Bowl because it produced an abundant root system and had a high resistance to drought and cold.[17]

Soil blowing was not the only agricultural problem Dust Bowl farmers faced. Just as had been true during the 1930s, the drought brought serious problems to the livestock industry. As early as the summer of 1952, the grass and feed supply was gone near Elkhart, Kansas. Two-thirds of the grass had dried up and there was no prospect for a feed crop that autumn. With rapidly declining feed supplies, the livestock situation became desperate. The seriousness of this problem was compounded because the grazing lands had been overstocked. Few cattlemen had the financial reserves to feed their stock through a prolonged drought, and they began to disperse their herds at distress prices. The cattle market became glutted, and the price dropped over 50 percent. In most cases, the returns were far below the amount cattlemen had invested in their livestock.[18]

Western cattlemen, while demanding complete freedom from government regulation during good times, have never been hesitant to seek federal aid when their pocketbooks became pinched. The 1950s were no exception. The Oklahoma Cattlemen's Association urged the federal government to buy 100 million pounds of beef at a minimum price of $.12 per pound on the hoof. The association also sought a renewal of the canning program initiated during the 1930s. Shading its real motives for beef price supports under a cloak of patriotism, the Oklahoma Cattlemen's Association recommended that government canned beef be distributed through the Office of Civil Defense "to guard against food shortage in the case of national emergency or disaster due to atomic attack." Cattlemen from the drought area also sought long term loans at low interest rates, low cost feed subsidized by the federal government if necessary, and emergency freight rates on feed for "bona fide" farmers and stockmen.[19]

T. L. Roach, president of the Texas and Southwestern Cattle Raisers Association, sent a telegram on June 24, 1953, to Congressman Clifford R. Hope, chairman of the House Agricultural Committee. In it Roach noted: "Cattlemen justifiably feel that the government should immediately do something to help the industry survive." He urged the government to sell cottonseed cake, corn, and other grains at reduced rates from its reserves and to grant credit for such purchases as well. This aid only seemed fair to the cattlemen, particularly "if the government intend[ed] continuing the use of food items for foreign-relief programs it could arrange to buy and can cattle, especially cows from the disaster area and use this canned meat in such relief programs."[20]

These demands did not go unheard. In late July 1953, the Commodity Credit Corporation began selling corn, wheat, oats, and cottonseed at below-market prices to help livestock producers maintain their basic herds. Additional aid was also forthcoming. As part of the new drought relief program, the federal government agreed to purchase 200 million pounds of beef or the equivalent of a million cattle for the American and Greek armies and for the school lunch program.[21]

The USDA also began cooperating with the drought states by providing funds to help pay railroad costs for transporting

hay into designated drought counties. In addition, the department gained the cooperation of several railroads in reducing
the freight rate 50 percent on incoming hay. By late October
1953, this program was in effective operation. Upon arrival
of the hay, farmers presented the freight agent with government certificates good for half of the freight bill and then paid
the remainder due from their own pockets. These certificates
were developed by the railroads and were sent in circular let-
ter form to the county agents. The certificates were then
mimeographed for distribution to the farmers. Once the certificates were signed by the county agent and the chairman of
the county drought committee, they were valid.[22]

The following year, in order to initiate the hay distribution
program, drought states governors had to request such aid
and sign a contract with the secretary of agriculture. These
contracts stipulated a specific amount of money to be made
available for the transportation of hay into designated
drought counties. Federal funds, however, could not be used
to pay more than half of the transportation cost. Under this
program the USDA did not actually acquire or distribute hay.
Instead, the department encouraged such acquisition through
"established channels" of trade. By so doing, these federal
funds helped to stimulate the local economy by giving business to local dealers. Farmers and ranchers obtained application forms for the hay program from the FHA and the county
ASCS committee office. These forms then required FHA
approval. When the farmer or stockman was certified to
receive assistance, he used the approved application form to
negotiate the cost of hay with the local dealer.[23]

The USDA also provided a feed grain program through the
FHA, which allowed farmers to purchase a sixty-day supply
of feed grain at a savings from the local price. Under this program, surplus stocks of corn, grain sorghum, oats, and barley
owned by the Commodity Credit Corporation could be
purchased through local dealers. The county ASCS committee extended negotiable certificates for $.60 per hundred-
weight for grain sold to eligible cattlemen. The Commodity

Credit Corporation then paid off these certificates from its surplus inventories. Money did not change hands, but grain did since the Commodity Credit Corporation used its reserves to compensate the dealers for selling their feed grains.[24]

In April 1955, the drought feed program was abandoned because the government expected great improvement in pastures and feed crop production. Although the program was reinstated for a brief time in several counties along the Kansas-Oklahoma line, applications were not accepted after June 15. The following autumn, the USDA allowed the FHA to extend additional credit for emergency loans beyond the December 31 termination date for its program. A total of 142 Dust Bowl counties were designated to receive this additional help.[25]

Special livestock loans were also available to Dust Bowl cattlemen who were unable to get loans from commercial banks or cooperative lending agencies. To qualify for a special loan, cattlemen had to have a good livestock record and have a reasonable chance to continue their operation with the aid of the loan. Livestock loans were made to help cattlemen meet unusual expenses for feed production; for replacing, renting, or repairing farm equipment; and for restocking herds. Loans were not made, to permit substantial expansion or for payment of old debts. These loans bore a 5 percent interest rate and were due in three years. If the situation warranted, the loans were renewable. The cattleman's available security served as collateral, and creditors were required to "stand by" and allow a "reasonable" part of the applicant's livestock profits to be used to repay the loan. This procedure allowed the cattlemen a chance to recover without threat of foreclosure as long as emergency drought conditions prevailed.[26]

The FHA also granted production emergency loans to Dust Bowl farmers who had experienced serious crop losses and who needed aid to continue their farming operations. These loans could be used for the purchase of feed, seed, fertilizer, replacement stock, and equipment; for essential home and farm operating expenses; or for replacing buildings and fences destroyed by dust storms. Production emergency loans bore a 3 percent interest rate and were secured with a chattel mortgage. Repayment was generally expected within five years, however, loans for the repair or improvement of real estate were due in ten years. By the end of August 1954, twenty-four Colorado counties, seventy-five Texas counties, thirty-seven Oklahoma counties, twenty-four New Mexico counties, and thirty Kansas counties had been designated emergency areas, thereby making these drought programs available to farmers and ranchers residing in them.[27]

During 1955, drought and soil blowing remained a menace, and the USDA modified its agricultural conservation program by providing a cost-sharing program for shifting crop lands out of production. Under this program, the USDA would pay up to 50 percent of the cost of seedbed preparation, fertilizer, and seed for establishing cover crops. Although the main purpose of this program was to provide protective cover, such crops had the added advantage of being good winter pasture.[28]

Two years later, as the drought was ending, the federal government also provided a deferred grazing program to help insure adequate protection for grazing lands. Under that program, farmers and ranchers in the drought area could receive federal dollars equal to the fair rental value of their land in exchange for not grazing livestock. Rental value was based on normal grazing capacity during periods of adequate precipitation. No payment would be made, however, if a cattleman removed livestock from one area and by so doing overgrazed another. The money available was supplementary to the agricultural conservation program, and it could not exceed $5,000 per person annually. Cattlemen could move livestock back onto the deferred grazing areas after November 1, but the grass had to be high enough so there would be no risk of overgrazing or wind erosion.[29]

In addition to proper soil conservation practices and government aid, irrigation also helped to stabilize agricultural lands and minimize the effects of drought in the 1950s Dust Bowl. Some farmers had begun irrigation along the Arkansas River in southwestern Kansas as early as 1880, but high costs and technological problems prevented extensive development of well irrigation until the 1930s. The drought of the 1930s convinced many Dust Bowl farmers that irrigation would stabilize or even increase crop production. By autumn 1934, farmers were irrigating in several Texas Panhandle and southern Plains counties, and Texas Tech University had a small irrigation well for demonstration purposes. Near Lockney, sixty-five farmers were irrigating, and in the Blackwater Valley near Muleshoe some 85 wells were pumping from 500 to 2,000 gallons per minute. By January 1935, nearly 300 irrigation wells watered over 35,000 acres in western Texas. Farmers in the southern Plains continued to expand irrigation during the droughty thirties and, by 1940, 2,180 wells irrigated over 250,000 acres.[30]

In southwestern Kansas, the sandy loam soil proved ideal for irrigation, and, by 1939, nearly 1,400 wells were in operation—about half had been dug since 1933. After the "dirty thirties," Kansas provides a good example of irrigation expansion in the Dust Bowl area. Not until the 1950s, however, did irrigation in the Kansas Plains expand to an exceptional degree. In 1954, 420,000 acres were irrigated (nearly 218,000 of this total were planted in sorghum); four years later 900,000 acres were irrigated. Several factors caused this expansion. First, farmers and scientists alike began to fear the return of a prolonged drought and Dust Bowl conditions, so farmers increasingly turned to irrigation to insure profitable harvests during the drought years. Second, irrigation systems became more convenient with the introduction of aluminum and gated pipe and center-pivot sprinklers. As a result, irrigation increased state-wide from 248,067 acres in 1949 to 537,566 acres in 1955 and to 1,020,573 acres in 1959. Much of that increase was in the old Dust Bowl portion of the state.[31]

Although irrigation permits a consistent annual growth of soil-holding crops, it changed the nature of Dust Bowl agriculture, and it created a potential problem—the exhaustion of the underground water supply. Today, cotton and corn grow where it could not possibly do so without irrigation. If the water table should ever become depleted, farmers will have little choice but to return to dry farming and make other adjustments to keep their lands from blowing during periods of drought. Nevertheless, while the water supply holds up, irrigation does help fight wind erosion, and it has truly made the Dust Bowl the land of the "underground rain."[32]

Although drought and erosion in the 1950s affected a larger area than during the 1930s, the Dust Bowl did not return to the distressing and tragic conditions of twenty years earlier. The conservation techniques implemented during the previous two decades prevented that. Dust Bowl farmers now understood the relationship between soil conservation and successful farming. Farmers no longer burnt their wheat stubble. Instead, they used one-way and chisel plows which tilled the soil but left the stubble on the surface. Contour plowing, strip cropping, and grazing management were now standard farming procedures. Furthermore, the Dust Bowl farmers did not have to contend with the financial problems of a depression, and so they were better able to survive the drought and to properly farm their blowing lands. The federal government once again provided funds for those farmers in financial difficulties so that they, too, could apply the proper conservation measures to their eroding fields.

During the 1930s, the Depression was already a major problem when the drought became severe. Many farmers were already bankrupt or had abandoned their farms when the dust began to blow. The economic situation was vastly different during the 1950s. In twenty-three Dust Bowl counties, farmers harvested 91 million bushels of wheat in 1952 for an average farm income of $54,000. In Greeley County, Kansas, farm production totalled $12 million for an average of $54,556 per farm. Although this was the last good harvest until the rains returned in the latter half of the decade, this income helped farmers to meet their extra tillage and operating expenses during the drought years. Even in 1954, the average farm income for that area was $12,000. At that same time, the national average income was less than $6,000. Sales tax collections are regarded as a reliable barometer for business conditions; in Kansas, these taxes climbed to record highs from 1953 into 1955. In addition, the continued development of large natural gas reserves in southwestern Kansas and the Oklahoma and Texas Panhandles gave royalty payments to landowners and further served to cushion the economic impact of the drought. Nor were these Dust Bowl farmers still reliant on a single crop for their livelihood. They had diversified—largely with sorghum crops—by the 1950s, and they no longer faced financial disaster if a wheat crop failed. As a result, farmers were able to stay on their land and practice the appropriate soil conservation techniques.[33]

Certainly, suitcase farming contributed to the wind erosion problem during the 1950s, but absentee owners probably did not contribute any more to causing dust storms than did negligent resident farmers. Suitcase farmers were responsible, though, for breaking a large amount of submarginal land— land that would not have blown during times of drought had it remained in grass. The AAA in the southeastern Colorado counties tried to prevent much of this nonresident expansion by imposing a penalty of $3 per acre for breaking native sod. But, the only way this money could be collected was to deduct it from the farmer's agricultural conservation program payment. If a suitcase farmer had not signed up for this program, the penalty could not be collected at all. With high wheat prices, good rains, and bumper crops during the 1940s, the wheat farmer—resident and nonresident alike—became independent of government control. Wheat farmers generally failed to readopt the anti-sodbreaking ordinances, and by 1952 only three conservation districts in eastern Colorado still had these restrictions in effect.[34]

In retrospect, the destruction of the native grass from the cumulative effects of drought, prairie fires, and overgrazing caused the dust storms of the nineteenth century; while the exposure of cultivated lands to drought and wind caused those of the twentieth. Certainly, the dust storms of the 1930s stand out because they occurred more frequently and with greater severity than before. Their cost in terms of damaged soil, ruined dreams, and human lives is, of course, incalculable. But this is not to say that the dust storms brought nothing but disaster to the southern Plains. Had the storms not rolled across the Plains during the 1930s, the work of the Soil Conservation Service would have been much smaller in scope. Indeed, Dust Bowl farmers would have considered such an agency unnecessary. Thus, the storms forced the farmers to contend with the problems of soil conservation. Although blowing dust plagued the southern Plains again during the 1950s and reminded careless farmers of the wind erosion hazard, the black blizzards did not recur. Significantly, Dust Bowl farmers profited from their knowledge of the past.[35]

Epilogue

Since the early 1950s, drought and blowing soil have continued to plague the Dust Bowl. In January 1965, the worst duster in a decade with seventy-five-mile-per-hour winds billowed soil thirty-one thousand feet high. Dust almost totally obscured the sun at 4:00 P.M. in the Dallas-Fort Worth vicinity—nearly three hundred miles from the center of the storm—and the wind sifted it as far east as Pennsylvania and West Virginia. By March, the old Dust Bowl had not received a good, general rain for eighteen months, and the wheat crop failed to grow sufficiently to anchor the soil. As a result, the dust blew. One Amarillo resident used a snow shovel three times that spring to scoop off the top layer of dust covering his lawn. Automatic sprinkler systems clogged throughout the city, and Amarillo residents began to claim "We don't go to the park. The park comes to us."[1]

Drought returned to the region a decade later. Kit Carson, Colorado, averaged less precipitation during the first six years of the 1970s than during the drought years of the 1930s. Here, the normal precipitation is 16.8 inches annually, but in 1975, the area received only 11 inches—35 percent below normal. Again the dust blew. One storm swept over Dodge City and reduced visibility to zero. Wind erosion became a serious problem in the spring of 1976, and drifting soil buried fence rows in Oklahoma. As the drought and soil blowing intensified, some farmers suffered total losses of their wheat crops.[2]

Wind erosion continued to be a problem into 1977 as alternate freezing and thawing, insufficient wheat growth, lack of crop residues, and poor farming methods contributed to soil blowing. On February 23, dust-laden winds reduced visibility to zero and paralyzed a twenty-county area in western Kansas. In some locations, traffic was halted along major highways and schools closed as forty-five to sixty-mile-per-hour winds churned the top soil into dust clouds that reached two miles high. The storm lasted twenty-four hours in Kit Carson County, Colorado, where it buried farm equipment, scoured paint off tractors, and covered fences. When the duster ended, 300,000 acres of cropland had been damaged and 40,000 acres of wheat had been destroyed. Vernon Haas, a U.S. soil conservationist at Burlington, Colorado, estimated that some fields suffered thirty years normal soil loss in that twenty-four hour period. Overall, wind erosion damaged 900,000 acres in a thirty-seven county area in eastern Colorado from November 1, 1976 to March 1, 1977; 242,000 acres of crops were lost, and an additional 2.2 million acres were in a condition to blow. Thirteen of these counties had been declared drought disaster areas and were therefore eligible for federal relief. In Kansas, Robert Griffin, an SCS conservationist at Salina, reported that the erosion situation was "potentially serious," as the wind scoured about 300,000 acres in a thirty-four-county area. Over 295,000 acres had been treated with emergency tillage, but an additional 3,390,550 acres were ready to blow. Similar conditions existed in Oklahoma, New Mexico, and Texas.[3]

Late spring rains broke the drought. Nevertheless, wind erosion during the 1977 blow season was the worst in the Dust Bowl since the 1950s. Colorado suffered the most severe loss with more than 2.5 million acres damaged, but the wind also removed top soil from 315,450 acres in Kansas, 206,840 acres in Oklahoma, 597,550 acres in New Mexico, and nearly 2.2 million acres in Texas.[4]

The sandy cotton lands of West Texas continued to blow during the winter. Early in the morning on December 16, a strong wind rose from the southwest in the Lubbock vicinity. Heavy dust was blowing at sunrise, and, by 10:00 A.M., as wind gusts reached sixty miles per hour, the fire department dispatcher reported "visibility zero and it might get worse." The storm blew all day. Motorists turned on headlights, and the air became dust-laden in homes and office buildings. Swimming pools turned to muddy water, and, on the Texas Tech University campus, the soil drifted curb deep.[5]

In short, the soil still blows in the region of the Dust Bowl when drought, wind, and inadequate vegetative cover provide the necessary ingredients for another dust storm. Nevertheless, if major dust storms similar to the black blizzards of the 1930s and to the dusters of the 1950s are not to return during periodic droughts, Dust Bowl farmers must continue to make major adjustments in their farming operations as changing conditions dictate. When the water table drops below levels where irrigation is no longer profitable, they must be quick to revert to wise dryland farming techniques. Furthermore, they must constantly realize the value of planting more drought-resistant crops, of diversification, and of reducing grazing on pasture lands during dry periods.[6]

Another Dust Bowl is not inevitable, but, given the right circumstances, it is possible.

Notes

CHAPTER 1

1. *Amarillo Daily News*, April 15, 1935; *Dodge City Daily Globe*, April 15, 1935; Guy Logsdon, "The Dust Bowl and the Migrant," *American Scene*, The Thomas Gilcrease Institute of American History and Art, 1971, pp. 1–2; Earl G. Brown et al., "Dust Storms and Their Possible Effect on Health," *Public Health Reports*, L (October 4, 1935), pp. 1373–74.

2. *Amarillo Daily News*, April 15, 1935; Logsdon, "The Dust Bowl," pp. 1–2; Brown, "Dust Storms and Their Possible Effect on Health," pp. 1373–74.

3. *Amarillo Daily News*, April 15, 1935; Logsdon, "The Dust Bowl," pp. 1–2; Robert R. Wilson and Ethel M. Sears, *History of Grant County, Kansas* (Wichita: Wichita Eagle Press, 1950), p. 184; Pauline W. Grey, "The Black Sunday of April 14, 1935," Pioneer Stories of Meade County, 1950, pp. 26–27, Kansas State Historical Society (hereafter cited as KSHS); Avis D. Carlson, "Dust," *New Republic*, 82 (May 1, 1935), p. 333.

4. "The Dust Bowl: Agricultural Problems and Solutions," U.S. Department of Agriculture, Office of Land Use Coordination, Editorial Reference Series No. 7, Washington, D.C., July 15, 1940, p. 1; Brown, "Dust Storms and Their Possible Effect on Health," p. 1370.

5. *Washington* (D.C.) *Evening Star*, April 15, 1935; Logsdon, "The Dust Bowl," pp. 2–3; Fred Floyd, "A History of the Dust Bowl," Ph.D. dissertation, University of Oklahoma, 1950, pp. 17–18.

6. William Van Royen, "Prehistoric Droughts in the Central Great Plains," *Geographical Review*, 27 (October 1937), pp. 637, 650; Harry E. Weakly, "Recurrence of Drought in the Great Plains During the Last 700 Years," *Agricultural Engineering,* 46 (February 1965), p. 85.

7. James C. Malin, "Dust Storms: Part One, 1850–1860," *Kansas Historical Quarterly*, 14 (May 1946), p. 129; Lela Barnes, "Journal of Isaac McCoy for the Exploring Expedition of 1830," *Kansas Historical Quarterly*, 5 (November 1936), pp. 365, 371.

8. Malin, "Dust Storms: Part One," pp. 133–35.

9. Ibid., p. 137; Snowden D. Flora, "Climate of Kansas," Report of the Kansas State Board of Agriculture, (June 1948), pp. 123–24.

10. Flora, "Climate of Kansas," p. 124; James C. Malin, "Dust Storms: Part Two, 1861–1880," Kansas Historical Quarterly, 14 (August 1946), pp. 268–69, 273, 276.

11. Malin, "Dust Storms: Part Two," pp. 280–87.

12. Ibid., pp. 283, 285.

13. Ibid., pp. 288-89.

14. Ibid., pp. 289-93.

15. *Monthly Weather Review*, January, 1883, p. 23, March, 1883, p. 73, and April, 1883, p. 97; James C. Malin, "Dust Storms: Part Three, 1881–1900," *Kansas Historical Quarterly*, 14 (November 1946), pp. 393, 396.

16. Malin, "Dust Storms: Part Three," p. 402; *Monthly Weather Review*, October, 1886, p. 296, January, 1887, p. 25, February, 1887, p. 59, April, 1887, pp. 1–4, and April, 1889, p. 94.

17. Flora, "Climate of Kansas," p. 125; *Salina* (Kansas) *Journal,* February 21, 1946.

18. *Kansas City Times*, April 19, 1935.

19. Angus McDonald, "Erosion and Its Control in Oklahoma Territory," United States Department of Agriculture, Miscellaneous Publication No. 301, March, 1938, p. 8; Sherwood B. Idso, "Dust Storms," *Scientific American*, 235 (October 1976), p. 108; Malin, " Dust Storms: Part Three," pp. 405–7; *Topeka Daily Capital,* April 7, 16, 18, and May 20, 1895.

20. Malin, "Dust Storms: Part Three," pp. 405–8; *Monthly Weather Review,* April, 1895, p. 130; James C. Malin, *Winter Wheat in the Golden Belt of Kansas* (Lawrence: University of Kansas Press, 1944), pp. 148–49.

21. *Country Gentleman*, 64 (March 16, 1899), p. 207.

22. Flora, "Climate of Kansas," p. 125; McDonald, "Erosion and Its Control in Oklahoma Territory," p. 8.

23. Flora, "Climate in Kansas," p. 126.

24. *Kansas City Star*, February 23 and June 5, 1913.

25. Ibid.

26. Ibid.

27. *Topeka Daily Capital,* July 10, 1913, and June 7, 1914.

28. *Topeka Daily Capital,* June 7, 1914; *Kansas City Star*, July 12, 1914; *Topeka Journal,* January 7, 1915.

29. McDonald, "Erosion and Its Control in the Oklahoma Territory," pp. 2–7.

CHAPTER 2

1. "The Dust Bowl: Agricultural Problems and Solutions," p. 3; R. S. Kifer and H. L. Stewart, *Farming Hazards in the Drought Area*, WPA, Division of Social Research, Research Monograph 16 (Washington, D.C.: n.p., 1938), p. 78; Arthur H. Joel, "Soil Conservation Reconnaissance Survey of The Southern Great Plains Area," U.S. Department of Agriculture, Technical Bulletin No. 556, 1937, p. 5.

2. "The Dust Bowl: Agricultural Problems and Solutions," p. 3; Kifer and Stewart, *Farming Hazards in the Drought Area,* pp. 79–80; Roy I. Kimmel, "A Long View of the Wind Erosion Problem," Kansas State Board of Agriculture Quarterly Report, (March 1938), pp. 87–90.

3. Joel, "Soil Conservation Reconnaissance Survey," pp. 8–9; James C. Malin, "Dust Storms are Normal," *University of Kansas Alumni Magazine,* 52 (March 1954), p. 4; C. Warren Thornwaite, "The Great Plains," *Migration and Economic Opportunity,* ed. Carter Goodrich (Philadelphia: University of Pennsylvania Press, 1936), p. 204; Ray Taffinder, secretary, Southwest Agricultural Association, Dalhart, Texas, to Senator Arthur Capper, May 7, 1937, Capper Papers, KSHS.

4. "The Dust Bowl: Agricultural Problems and Solutions," pp. 3–4; Joel, "Soil Conservation Reconnaissance Survey," p. 5; Flora, "Climate of Kansas," pp. 126–27; D. A. Savage, "Drought Survival of Native Grass Species in the Central and Southern Great Plains," U.S. Department of Agriculture, Technical Bulletin No. 549, 1937, pp. 4–5.

5. Thornwaite, "The Great Plains," pp. 207, 209–10.

6. Message from the President of the United States, "The Future of the Great Plains," House Doc. No. 144, 75th Cong., 1st sess., 1937, p. 4; U. S. Geological Survey, "Hydrology," *Twenty-Second Annual Report,* pt. 4 (Washington, D.C.: U.S. Government Printing Office, 1902), pp. 637–39, 654.

7. H. L. Stewart, "Changes on Wheat Farms in Southwestern Kansas, 1931–37," U. S. Department of Agriculture, Farm Management Reports, No. 7, Washington, D.C., June, 1940, p. 4.

8. "The Dust Bowl: Agricultural Problems and Solutions," pp. 8–10; Hugh Hammond Bennett, *Soil Conservation* (New York: McGraw-Hill, 1939), pp. 733–36.

9. "The Dust Bowl: Agricultural Problems and Solutions," pp. 9–11.

10. Malin, *Winter Wheat in the Golden Belt of Kansas,* p. 104; Walter Prescott Webb, *The Great Plains in Transition* (New York: Grosset and Dunlap, 1931), p. 372; John D. Sjo, "Technology: Its Effects on the Wheat Industry," Ph.D. dissertation, Michigan State University, 1960, p. 61.

11. Wayne D. Rasmussen, "The Impact of Technological Change on American Agriculture," *Journal of Economic History,* 20 (December 1962), pp. 579–80; Donald P. Green, "Prairie Agricultural Technology, 1860–1900," Ph.D. dissertation, Indiana University, 1957, pp. 174, 210–12; Malin, *Winter Wheat,* pp. 60, 210; Patricia M. Bourne and A. Bower Sagerser, "Background Notes on the Bourne Lister Cultivator," *Kansas Historical Quarterly*, 20 (August 1952), p. 183.

12. Carl Frederick Kraenzel, *The Great Plains* (Norman: University of Oklahoma Press, 1969), pp. 135, 302; Malin, *Winter Wheat,* pp. 61–65.

13. Kraenzel, *The Great Plains,* p. 135; "The Dust Bowl: Agricultural Problems and Solutions," p. 12.

14. Ibid.; Lawrence Svobida, *An Empire of Dust* (Caldwell, Idaho: Caston Printers, 1940), pp. 39–40; Glenn K. Rule, "Crops Against the Wind on the Southern Great Plains," U.S. Department of Agriculture, Farmers' Bulletin No. 1833, 1939, pp. 5–6; Bennett, *Soil Conservation*, p. 737; *The Future of the Great Plains,* p. 4; "A Report of the Soil Conservation Service to the Secretary of Agriculture on Problems of the Southern Great Plains and a Conservation Program for the Region," April, 1954, p. 11, National Agricultural Library (hereafter cited as NAL).

15. Stewart, "Changes on Wheat Farms in Southwestern Kansas," p. 5; Vance Johnson, *Heaven's Tableland* (New York: Farrar, Straus, 1947), pp. 129, 131, 147.

16. Stewart, "Changes on Wheat Farms in Southwestern Kansas," p. 5; Frances McNeill Alsup, "A History of the Panhandle of Texas," Master's thesis, University of Southern California, 1943, p. 191; Paul B. Sears, *Deserts on the March* (Norman: University of Oklahoma Press, 1935), p. 76; Svobida, *An Empire of Dust*, p. 33.

17. Rule, "Crops Against the Wind on the Southern Great Plains," p. 5; Earl H. Bell, *Culture of a Contemporary Rural Community: Sublette, Kansas,* Rural Life Studies: 2, U.S. Department of

Agriculture, Bureau of Agricultural Economics, September 1942, p. 14.

18. Garry L. Nall, "Specialization and Expansion: Panhandle Farming in the 1920's," *Panhandle-Plains Historical Review*, 47 (1974), pp. 47, 63; Sears, *Deserts on the March*, p. 121.

19. W. E. Grimes, "The Effect of Improved Machinery and Production Methods on the Organization of Farms in the Hard Winter Wheat Belt," *Journal of Farm Economics*, 10 (1928), p. 226; Stewart, "Changes on Wheat Farms in Southwestern Kansas," p. 6; Paul G. Beck and M. C. Forster, "Six Rural Problem Areas, Relief-Resources-Rehabilitation," FERA Research Monograph 1, Washington, D.C., 1935, p. 18; Paul Bonnifield, "The Oklahoma Panhandle's Agriculture to 1930," *Red River Valley Historical Review*, 3 (Winter 1978), p. 75.

20. "The Dust Bowl: Agricultural Problems and Solutions," p. 13.

21. Ibid., p. 14.

22. Leslie Hewes, *The Suitcase Farming Frontier* (Lincoln: University of Nebraska Press, 1973), pp. 13–15; Thornwaite, "The Great Plains," pp. 215–16; Bennett, *Soil Conservation*, p. 737; *The Future of the Great Plains*, p. 4.

23. McDonald, "Erosion and Its Control in Oklahoma Territory," pp. 4–6; H. H. Finnell, "Prevention and Control of Wind Erosion of the High Plains Soils in the Panhandle Area," p. 9 (mimeographed), Soil Conservation Service Publication, NARG 114; B. W. McGinnis, "Utilization of Crop Residues to Reduce Wind Erosion," *The Land Today and Tomorrow*, 2 (April 1935), p. 12; "The Report of the Eleventh Conference of the Regional Advisory Committee on Land Use in the Southern Great Plains Area," Colorado Springs, Colorado, April 19–20, 1937, pp. 1–2, 5–6, Office of the Secretary of Agriculture, General Correspondence, 1937, Drought File, NARG 16; R. I. Throckmorton, "A Soil Conservation Program for Kansas," Thirty-First Biennial Report of the Kansas State Board of Agriculture (1937–1938), p. 39; C. C. Isley, "Will the Dust Bowl Return?" *The Northwestern Miller*, 224 (November 20, 1945), p. 35; Ben Hibbs, "Reaping the Wind," *Country Gentleman*, 104 (May 1934), pp. 45, 48; Thornwaite, "The Great Plains," p. 237.

24. Ira Wolfert, *An Epidemic of Genius* (New York: Simon & Schuster, 1960), p. 79; *The Future of the Great Plains*, pp. 27, 29; Joel, "Soil Conservation Reconnaissance Survey of the Southern Great Plains Area," pp. 15, 22, 25; "A Survey of Baca County Colorado," July 14, 1934, Bureau of Agricultural Economics, Rural

Problem Area Reports, pp. 5, 43, NARG 83; "A Survey of Meade County Kansas," August 11, 1934, Bureau of Agricultural Economics, Rural Problem Area Reports, p. 39, NARG 83; H. V. Geib, "Report of the Wind Erosion Survey in the Region of the Oklahoma Panhandle and Adjacent Territory," p. 3, Office of the Secretary of Agriculture, General Correspondence, 1933, Drought File, NARG 16; A. W. Zingg, "Speculations on Climate as a Factor in the Wind Erosion Problem of the Great Plains," *Transactions of the Kansas Academy of Science,* 56 (September 1953), pp. 371–77; Bennett, *Soil Conservation,* p. 739; Thornwaite, "The Great Plains," p. 238; John C. Hoyt, "Droughts of 1936 with Discussion on the Significance of Drought in Relation to Climate," U.S. Department of the Interior, Washington, D.C., 1938, p. 28.

25. Rule, "Crops Against the Wind on the Southern Great Plains," p. 32; H. T. U. Smith, *Geological Studies in Southwestern Kansas,* Vol. 41 (September 15, 1940), pp. 178–79; Geib, "Report of the Wind Erosion Survey in the Region of the Oklahoma Panhandle and Adjacent Territory," p. 4, NARG 16; *Yearbook of Agriculture, 1935* (Washington, D.C.: U.S. Government Printing Office, 1935), pp. 15–16; "A Report of the Soil Conservation Service to the Secretary of Agriculture on Problems of the Southern Great Plains and a Conservation Program for the Region," p. 3, NAL; *Monthly Weather Review,* 63 (February 1935), p. 53.

26. Grimes, "The Effect of Improved Machinery and Production Methods," pp. 226–28.

27. "The Dust Bowl: Agricultural Problems and Solutions," p. 15; J. E. Weaver and F. W. Albertson, *Grasslands of the Great Plains* (Lincoln, Nebr.: Johnsen Publishing Co., n.d.), p. 92.

28. "Dust Storms of the Vicinity of Amarillo, Texas," data taken from the Records of the United States Weather Bureau by H. T. Coleman, May 10, 1936, McCarty Collection, Amarillo Public Library (hereafter cited as APL); *Monthly Weather Review,* 59 (January 1931), p. 30; Bennett, *Soil Conservation,* p. 121; B. Ashton Keith, "A Suggested Classification of Great Plains Dust Storms," *Transactions of the Kansas Academy of Science,* 47 (1944–45), pp. 106–9.

CHAPTER 3

1. *Amarillo Globe,* January 22 and April 24, 1932; Floyd, "A History of the Dust Bowl," p. 44; "Dust Storms in the Vicinity of Amarillo, Texas," data taken from records of the United States Weather Bureau, Amarillo, Texas, by H. T. Coleman, May 10, 1936, McCarty Collection, APL; W. G. Deloach Papers, Diary 6, p. 146, Southwest Collection, Texas Tech University.

2. *Amarillo Globe,* April 30, 1933; Beeson Museum Dust Bowl Scrapbook, KSHS; *Dalhart Texan,* January 10, 1934; *Monthly Weather Review,* 62 (January 1934), pp. 12–13; *Elkhart* (Kansas) *Tri-State News,* May 11, 1933. By August 10, 1933, Goodwell, Oklahoma, had experienced more than thirty dust storms, about ten times the number considered normal for that period.

3. *Monthly Weather Review,* 62 (January 1934), pp. 12–13; *Monthly Weather Review,* 62 (May 1934), p. 162; Floyd, "A History of the Dust Bowl," pp. 66–67; *Dodge City Daily Globe,* May 5, 1934; *Monthly Weather Review,* 63 (February 1935), p. 53; Logsdon, "The Dust Bowl," p. 2; *Monthly Weather Review,* 62 (May 1934), p. 156; "A Report of the Soil Conservation Service to the Secretary of Agriculture on Problems of the Southern Great Plains and a Conservation Program for the Region," p. iii, NAL; *New York Times,* March 31, 1935; *Amarillo Globe,* May 11, 1935.

4. *Amarillo Globe,* March 26, April 11 and 12, 1934; Don Eddy, "Up from Dust," *American Magazine,* 129 (April 1940), p. 89.

5. *Dodge City Daily Globe,* May 5, 1934.

6. *Amarillo Globe,* April 18, May 3 and 11, 1934; *Dodge City Daily Globe,* May 5, 1934.

7. Unknown Kansas newspaper clipping, February 22, 1935; *Dodge City Daily Globe,* February 22, 1935; *Amarillo Globe,* February 22 and 25, 1935.

8. *Monthly Weather Review,* 63 (March 1935), p. 144; Ibid., (May 1935), p. 175.

9. *Amarillo Globe,* March 4, 1935; W. G. Deloach Papers, Diary 6, p. 383; *Kansas City Times,* March 20, 1935; Weaver and Albertson, *Grasslands of the Great Plains,* pp. 91–92; Floyd, "A History of the Dust Bowl," pp. 89, 91.

10. *Amarillo Globe,* March 20 and 26, 1935; *Dodge City Daily Globe,* March 20, 1935; *Fort Worth Star-Telegram* (Evening), March 17 and 20, 1935.

11. *Amarillo Daily News,* April 10, 1935; Floyd, "A History of the Dust Bowl," p. 93.

12. *Garden City* (Kansas) *Daily Telegram,* April 10, 1935; Beeson Museum Scrapbook, KSHS; *Fort Worth Star-Telegram* (Evening), April 11, 1935; *Dodge City Daily Globe,* April 19, 1935.

13. "Dust and the Nation's Bread-Basket," *Literary Digest,* 119 (April 20, 1935), p. 10; *Fort Worth Star-Telegram,* May 3, 1935; *Dodge City Daily Globe,* May 3, 1935.

14. *Topeka Journal,* August 16, 1935; The Dodge City park crew scooped about 130 tons of dirt from the swimming pool in the spring. See Floyd, "A History of the Dust Bowl," p. 105.

15. *Pueblo* (Colorado) *Indicator,* February 29, 1936; *Monthly Weather Review,* 64 (June 1936), p. 197.

16. Ibid.; "News From the Dust Bowl," Clifford R. Hope Papers, KSHS.

17. Floyd, "A History of the Dust Bowl," pp. 150–52; *Monthly Weather Review,* 65 (April 1937), pp. 151–52.

18. Ibid.; E. F. Chilcott to Dr. Leighty, Woodward, Oklahoma, February 8, 1937, Clifford R. Hope Papers, KSHS; *Amarillo Globe,* February 17 and 21, 1937.

19. *Monthly Weather Review,* 65 (April 1937), p. 152; *Denver Post,* March 9, 1937.

20. *Monthly Weather Review,* 65 (April 1937), p. 152.

21. Ibid., 66 (January 1938), p. 9; *Amarillo Sunday News and Globe,* May 23, 1937; Floyd, "A History of the Dust Bowl," p. 154.

22. *Monthly Weather Review,* 66 (January 1938), p. 10.

23. Ibid., p. 11.

24. Ibid., pp. 11–12.

25. Ibid., 67 (January 1939), p. 12.

26. Ibid., pp. 12-13.

27. Ibid., p. 13.

28. Ibid., pp. 13–14; *Amarillo Globe,* January 2, 1939.

29. *Monthly Weather Review,* 67 (January 1939), p. 14.

30. Ibid., pp. 14–15.

31. Ibid., pp. 446–47.

32. Ibid., 67 (December 1939), pp. 446–47; Floyd, "A History of the Dust Bowl," p. 174.

33. *Monthly Weather Review,* 67 (December 1939), pp. 447–51.

34. "A Report of the Soil Conservation Service to the Secretary of Agriculture on the Problems of the Southern Great Plains and a Conservation Program for the Region," p. iii, NAL; C.W.A. Subject File, Documentary Resources, Colorado State Historical Society; A. W. Malone, "Desert Ahead!" *New Outlook,* 164 (August 1934), p. 14. From January 1933 to February 1936, Amarillo experienced 217 days of dust storms during which visibility was limited to six miles or less. See Colman, "Dust Storms of the Vicinity of Amarillo, Texas," p. 10.

CHAPTER 4

1. Beeson Museum Scrapbook, KSHS; Floyd, "A History of the Dust Bowl," p. 110; Carlson, "Dust," p. 333; Svobida, *An Empire of Dust,* p. 104; "Dust: More Storms Wreck Destruction in the Southwest," *Newsweek,* (April 20, 1935), p. 11; Stanley Vestal, *Short Grass Country* (New York: Duell, Sloan & Pearce, 1941), p. 203.

2. Dust Bowl manuscripts, Campbell Collection, Western History Collections, University of Oklahoma; *Denver Post,* April 15,

1935; Flora, "Climate of Kansas," pp. 271–72; Dawson Scrapbook, Vol. 57, p. 221, Colorado State Historical Society; *Fort Worth Star-Telegram,* March 25, 1935; *Elkhart* (Kansas) *Tri-State News,* April 20, 1933.

3. *Springfield* (Colorado) *Democrat-Herald,* January 10 and March 21, 1935; *Elkhart* (Kansas) *Tri-State News,* January 17 and 31, 1935.

4. *Topeka Capital,* March 17, 1935; Svobida, *An Empire of Dust,* pp. 92, 102; *The History of Smith Center, Kansas* (N.P.: N.P., 1971), p. 15, KSHS.

5. Brown, "Dust Storms and Their Possible Effect on Health," pp. 378–83; Floyd, "A History of the Dust Bowl," p. 111; Wilson and Sears, *A History of Grant County, Kansas,* p. 184; Dust Bowl Scrapbook, KSHS; *Liberal* (Kansas) *News,* May 29, 1935. The hospitals were at Walsh and Springfield, Colorado; Johnson, Cimarron, and Ulysses, Kansas; and Stratford, Texas.

6. *Rocky Mountain* (Denver) *News,* March 23, 1935; *Amarillo Daily News,* April 29 and 30, 1935; "Dust and the Nation's Bread-Basket," p. 10.

7. *Dodge City Daily Globe,* May 11, 1935.

8. "Dust-Storm's Aftermath," *Literary Digest,* 120 (November 2, 1935), p. 15; Brown, "Dust Storms and Their Possible Effect on Health," pp. 1380–83.

9. *Topeka Capital,* March 17, 1935; *Amarillo Sunday News and Globe,* March 17, April 27, and May 1, 1935.

10. *Amarillo Daily News,* May 1, 2 and 3, 1935.

11. Beeson Museum Dust Bowl Scrapbook, KSHS.

12. *Dodge City Daily Globe,* April 15, 1935.

13. Ibid.; Beeson Museum Dust Bowl Scrapbook, KSHS.

14. *Topeka Journal,* March 20, 1935; David Nail, *One Short Sleep Past: A Profile of Amarillo in the Thirties* (Canyon, Tex.: Staked Plains Press, 1973), pp. 113–14; *Amarillo Globe,* April 10, 1935.

15. Dust Bowl Manuscripts, Campbell Collection, Western History Collections, University of Oklahoma; *Dodge City Daily Globe,* March 29, 1935; *Springfield* (Colorado) *Democrat-Herald,* May 2, 1935, and February 17, 1936; *Elkhart* (Kansas) *Tri-State News,* March 12, 1936.

16. Ruby Winona Adams, "Social Behavior in a Drought Stricken Panhandle Community," Master's thesis, University of Texas, 1939, p. 35; Mark A. Dawber, "Churches in the Dust Bowl," *Missionary Review of the World,* 62 (September 1939), pp. 394, 397.

17. Beeson Museum Dust Bowl Scrapbook, KSHS: Josephine Strode, "Kansas Grit," *Survey,* 67 (August 1936), p. 230; Nail, *One Short Sleep Past,* p. 116; *Amarillo Globe,* April 12, 1935.

18. Vestal, *Short Grass Country,* pp. 205–6; Strode, "Kansas Grit," p. 231; "Drought Strikes the Great Plains Again," *Business Week* (November 18, 1939), p. 16.

19. *Fort Worth Star-Telegram,* March 20, 1935.

20. *Dalhart Texan,* March 11, 1935; Last Man's Club Folder, McCarty Collection, APL; Floyd, "A History of the Dust Bowl," p. 120.

21. John Dawson interview with John L. McCarty, February 12, 1972, in the possession of Mrs. Evelyn Jeanne Claytor, Amarillo, Texas (hereafter cited as Claytor MSs).

22. *Dalhart Texan,* October 17, 1941; John L. McCarty to Ed Bishop, Dalhart, Texas, no date, Claytor MSs.

23. *Muleshoe* (Texas) *Journal,* March 7, 1935.

24. Walter Davenport, "Land Where Our Children Die," *Colliers* (September 18, 1937), p. 12; "Thou Shalt Not Bear False Witness," McCarty Collection, APL.

25. Ibid.; *Southwest* (Liberal, Kansas) *Tribune,* September 23, 1937; John L. McCarty to Ray Millman, Editor, *Southwest* (Liberal, Kansas) *Tribune,* September 29, 1937, McCarty Collection, APL.

26. McCarty Sandstorm Tribute Scrapbook, Claytor MSs.

27. Logsdon, "The Dust Bowl," p. 9; Mark Van Doren, "Further Documents," *Nation,* 142 (June 10, 1936), p. 753; "Dust Storm Film," *Literary Digest,* 121 (May 16, 1936), pp. 22–23; Floyd, "A History of the Dust Bowl," p. 159; *Amarillo Globe,* May 11, 1936.

28. *Amarillo Sunday News and Globe,* May 31, 1936.

29. *Amarillo Globe,* June 9, 1936; *Fort Worth Star-Telegram* (evening), June 9, 1936; "Federal Movie Furor," *Business Week* (July 11, 1936), p. 14. The film was only 2,700 feet and lasted twenty-eight minutes. See *Time,* May 25, 1936, pp. 47–48.

30. *Amarillo Globe,* May 31 and July 3, 1936; Isley, "Will the Dust Bowl Return," p. 35.

31. *Amarillo Globe,* June 18, 1937.

32. Alsup, "A History of the Panhandle of Texas," pp. 221, 240; *Dalhart Texan,* June 12, 1937; *Fort Worth Star-Telegram* (morning), June 18, 1937.

33. *Springfield* (Colorado) *Democrat-Herald,* March 28, 1935; *Fort Worth Star-Telegram,* June 28, 1935.

34. Logsdon, "The Dust Bowl," pp. 6, 16; *Amarillo Globe,* June 18, 1937; *Kansas Farmer,* April 27, 1935; Adams, "Social Behavior," pp. 66, 69, 82.

35. *Topeka Capital,* February 28, 1935; Ada Buell Norris, "Black Blizzard," *Kansas Magazine* (1941), p. 104.

36. Caroline A. Henderson, "Letters from the Dust Bowl," *Atlantic,* 157 (May 1936), p. 543.

CHAPTER 5

1. Bennett, *Soil Conservation,* p. 55; E. E. Free and J. M. Westgate, "The Control of Blowing Soils," *Farmers' Bulletin No. 421,* 1910, p. 3; Hugh Hammond Bennett, "Facing the Erosion Problem," *Science,* 81 (April 5, 1935), p. 323; Ivan Ray Tannehill, *Drought: Its Causes and Effects* (Princeton: Princeton University Press, 1947), pp. 12, 41.

2. McDonald, "Erosion and Its Control in Oklahoma Territory," pp. 25–28; *Topeka Daily Capital,* July 10, 1913; Free and Westgate, "The Control of Blowing Soils," p. 11; Victor C. Seibert, "A New Menace to the Middle West: The Dust Storms," *The Aerend,* 8 (Fall 1937), p. 213.

3. E. Morgan Williams, "The One-Way Disc Plow: Its Historical Development and Economic Role," Master's thesis, University of Kansas, 1962; E. F. Chilcott, "Preventing Soil Blowing on the Southern Great Plains," U.S. Department of Agricultures Farmers' Bulletin No. 1771, 1937, pp. 3–4; L. C. Aicher, "Curbing the Wind," Twenty-Ninth Biennial Report of the Kansas State Board of Agriculture, (1933–1934), p. 68; *Dodge City Daily Globe,* April 14 and 22, 1935; *Abilene* (Kansas) *Daily Reflector,* March 26, 1935; J. S. Plough, "Out of the Dust," *Christian Century,* 52 (May 22, 1935), p. 692; Robert S. Field to John L. McCarty, April 4, 1935, Sandstorm Tribute Scrapbook, Claytor MSs; Free and Westgate, "The Control of Blowing Soils," p. 3; Robert S. Field to Clifford R. Hope, February 17, 1937, Hope Papers, KSHS.

4. N. P. Woodruff, W. S. Chepil, and R. D. Lynch, "Emergency Chiseling to Control Wind Erosion," Kansas State Agricultural Experiment Station, 1957, Technical Bulletin No. 90, p. 1; Svobida, *An Empire of Dust,* pp. 62, 199; Wolfert, *An Epidemic of Genius,* pp. 73–74; Chilcott, "Preventing Soil Blowing on the Southern Great Plains," pp. 3–12; L. C. Aicher, "The Fort Hays Daming Attachment for Listers," Thirtieth Biennial Report of the Kansas State Board of Agriculture, (1935–1936), p. 82; *Topeka Daily Capital,* August 30, 1936; Aicher, "Curbing the Wind," p. 71; Hibbs, "Reaping the Wind," p. 48. For a description of a homemade daming lister built by C. T. Peacock of Lincoln County, Colorado, see Forrest Albert Young, "The Repercussions on the Economic System of the Great Plains Region of Kansas of the Mechanization of Agriculture," Ph.D. dissertation, State University of Iowa, 1938, pp. 93–98.

5. *Dodge City Daily Globe,* March 28 and April 2, 11, and 12, 1935; *Kansas City Times,* March 14, 1936; *Topeka Capital,* April 5, 1935; *The Future of the Great Plains,* p. 3; *Amarillo Daily News,*

March 17 and April 11, 12, and 17, 1935. Blank listing means plowing listed furrows solidly across the entire field.

6. "The First Five Years of the Regional Agricultural Council for
the Southern Great Plains States," prepared by the Office of Land
Use Coordination, Amarillo, Texas, c. 1941, pp. 16–18, Records of
the Soil Conservation Service, Prairie States Forestry Project,
NARG 114.

7. Hibbs, "Reaping the Wind," p. 48; Chilcott, "Preventing Soil
Blowing on the Southern Great Plains," pp. 4–7.

8. Chilcott, "Preventing Soil Blowing on the Southern Great
Plains," pp. 7–10; Svobida, *An Empire of Dust,* p. 202.

9. Kifer and Stewart, *Farming Hazards in the Drought Area,* p. 80;
H. H. Bennett, *Handbook of Soil and Water Conservation Practices for
the Wind Erosion Area* (Washington, D.C.: U.S. Department of
Agriculture, 1936), pp. 24–25; Margaret Bourke-White, "Dust
Changes America," *Nation,* 140 (May 22, 1935), p. 597; *Amarillo Sunday News and Globe,* September 30, 1934, and January 20,
1935.

10. H. H. Bennett, "Memorandum to the Secretary of Agriculture Regarding A Plan for Control of Wind Erosion in the Region of
Southwestern Kansas, Western Oklahoma, and Northwestern
Texas in Connection with Agricultural Relief," August 22, 1933,
Office of the Secretary of Agriculture, Drought File, NARG 16;
Russell Hatfield, "People on the Plains," *Kansas Water News,* 13
(1970), pp. 11–12; Ben Hibbs, "Dust Bowl," *Country Gentleman,*
210 (March 1936), p. 85; *Amarillo Sunday News and Globe,* August 5
and 12, September 9, October 7 and 9, November 11, 1934, and
June 23, and July 14, 1935, and August 9, 1936; Glenn K. Rule,
"Land Facts on the Southern Great Plains," U.S. Department of
Agriculture, Miscellaneous Publication No. 334, 1939, p. 7.

11. Throckmorton, "A Soil Conservation Program for Kansas,"
pp. 47–48; Rule, "Crops Against the Wind on the Southern Great
Plains," p. 19; "Emergency Tillage Operations and Results," pp.
1–2, Soil Conservation Service Drought File, 1936–1937, NARG
114; H. H. Finnell, "The Progress Made in Wind Erosion Control
in the Southern High Plains Region," for the Seventh Southwestern
Soil and Water Conservation Conference, Stillwater, Oklahoma,
July 7–8, 1936, p. 5, Soil Conservation Service Drought File,
1936–1937, NARG 114; *Amarillo Sunday News and Globe,* July 19
and August 30, 1936.

12. Hatfield, "People on the Plains," pp. 12–15; H. H. Finnell,
"Annual Report, 1939–1940," p. 9, Region 6, Soil Conservation
Service Publications, NARG 114; Clifford R. Hope to E. C. Sum-
mers, June 1, 1937, Hope Papers, KSHS; "Soil Conservation Dis-

tricts for Erosion Control," United States Department of Agriculture, Miscellaneous Publication No. 293, 1937, pp. 10–14.

13. Clifford R. Hope to Ray Johnson, June 2, 1937, Hope Papers, KSHS; Kansas Legislative Council, Research Department, "Soil Drifting," Publication No. 43, November 12, 1936, pp. 8–9; *Amarillo Daily News,* May 22, 1935; C. W. Humble, "Nine Counties Have Organized Wind Erosion Districts," in "High Plains Conservationist" (mimeographed), Soil Conservation Service Region 6, Amarillo (March 1936), pp. 3–4, SCS Publications, NARG 114. By March 1936, nine Texas Panhandle counties had organized wind erosion districts. They were Dallam, Hartley, Oldham, Deaf Smith, Sherman, Moore, Lipscomb, Hanford, and Ochiltree. Other counties were preparing to organize similar districts.

14. "Soil Drifting," p. 2.

15. Ibid., pp. 2–3.

16. Ibid.; *Kansas ex. rel. Perkins* v. *Hardwick et al.,* 144 Kan. 3.

17. Ibid.; Kansas. *Session Laws, 1937,* ch. 189; *Elkhart* (Kansas) *Tri-State News,* April 1, 1937.

18. Hatfield, "People on the Plains," p. 14; "Dust-Bowl into Grazing-Land," *Literary Digest,* 121 (March 7, 1936), p. 9.

19. H. H. Finnell, "Appraisal of South Plains Agriculture Conditions with Recommendations of a Permanent Program to the Drought Situation," August 17, 1936, pp. 8–9, Soil Conservation Service Drought File, 1936–1937, NARG 114; *Amarillo Globe,* December 18, 1932, and September 1, 1935.

20. "Emergency Tillage Operations and Results," pp. 1–2, NARG 114; "Program for Rural Reconstruction," pp. 4–5, NARG 114; Rule, "Crops Against the Wind on the Southern Great Plains," p. 21; *Amarillo Sunday News and Globe,* February 3, 1935, and March 8, 1936.

21. "SCS Work Tested," Soil Conservation Service Drought File, 1936–1937, NARG 114; "Colorado Conservancy," (mimeographed), Colorado Springs, Colorado (May and June, 1936), p. 1, Soil Conservation Service Publications, NARG 114; H. H. Finnell, "Water Management," Soil Conservation Service Publications, NARG 114; *Amarillo Sunday News and Globe,* August 9, 1936.

22. Bennett, *Soil Conservation,* pp. 346, 360–62; Rule, "Crops Against the Wind on the Southern Great Plains," pp. 30, 33–37; Bennett, *Handbook of Soil and Water Conservation,* pp. 29, 31–32; R. R. Hinde, "Strip Cropping of Corn, Cotton and Beans," in "High Plains Conservationist," (mimeographed), Soil Conservation Service Region 6, Amarillo (March 1936), p. 8, Soil Conservation Service Publications, NARG 114.

23. Hinde, "Strip Cropping of Corn, Cotton and Beans," p. 8, NARG 114; Chilcott, "Preventing Soil Blowing on the Southern Great Plains," p. 25.

24. Joel, "Soil Conservation Reconnaissance Survey of the Southern Great Plains Area," pp. 6, 8; Rule, " "Crops Against the Wind on the Southern Great Plains," pp. 39–40; J. C. Whitfield, "Wind Erosion Endangering Colorado Vegetation," *The Land Today and Tomorrow,* 1 (December 1934), pp. 27–28; "The Recent Destructive Dust Cloud," *Science,* 97 (May 25, 1934), p. 473; H. V. Woodman, "Pasture Development in Texas," *The Land Today and Tomorrow,* 2 (March 1935), p. 7; "The Report of the Eleventh Conference of the Regional Advisory Committee on Land Use in the Southern Great Plains Area," p. 8, NARG 16; Bennett, *Handbook of Soil and Water Conservation,* pp. 39–40; "The Drought Situation," p. 4, Office of the Secretary of Agriculture, General Correspondence, Drought File, 1935, NARG 16.

25. "The Use of Contour Furrows and Related Structures on Pasture and Range Land," prepared by the Section of Agronomy and Range Management, Region 6, Amarillo, Texas, for the Soil Conservation Conference on Agronomy and Range Management, Denver, Colorado, January 13–16, 1937, pp. 1–6, (mimeographed), Soil Conservation Service Publications, NARG 114; Eugene C. Buie, "Contour Furrowing in Region 6," pp. 1–2, (mimeographed), Soil Conservation Service Publications, NARG 114.

26. "The Use of Contour Furrows and Related Structures on Pastures and Range Land," pp. 12–23, NARG 114; Buie, "Contour Furrowing in Region 6," p. 3, NARG 114; "Annual Report," A.S.A.E. Subcommittee on Contour Farming, 1939–1940, Soil Conservation Service Publications, NARG 114; Woodman, "Pasture Development in Texas," p. 11.

27. B. W. McGinnis, "Erosion and Its Control on the Southern High Plains," p. 16, (mimeographed), Soil Conservation Service Publications, NARG 114; Sydney H. Watson, "Natural and Artificial Revegetation in the Southern Great Plains," pp. 1–2, (mimeographed), Soil Conservation Service Publications, NARG 114; F. S. Reynolds, "Seeding Eroded and Cultivated Land to Native Grasses on Dalhart Project," in "High Plains Conservationist" (mimeographed), Soil Conservation Service Region 6, Amarillo (March 1936), pp. 10–11, Soil Conservation Service Publications, NARG 114.

28. Watson, "Natural and Artificial Revegetation in the Southern Great Plains," pp. 2–5, NARG 114; "Program for Rural Reconstruction," pp. 3–4, NARG 114.

29. Watson, "Natural and Artificial Revegetation," p. 4, NARG 114; Rule, "Crops Against the Wind on the Southern Great Plains," pp. 45–46; Chilcott, "Preventing Soil Blowing on the Southern Great Plains," p. 27.

30. Weaver and Albertson, *Grasslands of the Great Plains,* p. 92; Kimmell, "A Long View of the Wind Erosion Problem," p. 86.

31. Charles J. Whitfield, "Sand-Dune Reclamation in the Southern Great Plains," U.S. Department of Agriculture, Farmers' Bulletin No. 1825, 1939, pp. 1–2, 7–9; *Amarillo Sunday News and Globe,* November 17, 1935; Charles J. Whitfield, "Sand Dunes in the Great Plains," *Soil Conservation,* 2 (March 1937), p. 209.

32. Memorandum to Paul H. Appleby, Chairman, Land Policy Committee from (?), July 11, 1936, SCS, General File, Drought, 1936–1937, NARG 114; W. R. Watson, "Colorado Drought Relief Project," c. February 1937, SCS, General File, Drought, 1936–1937, NARG 114.

33. Ibid.; "Drought," SCS, General File, Drought, 1936–1937, NARG 114.

34. *Belleville* (Kansas) *Telescope,* June 6, 1935.

35. *Elkhart* (Kansas) *Tri-State News,* April 29, 1937; Malone, "Desert Ahead!" p. 14; Walter A. Huxman to H. A. Kinney, May 4, 1937, Huxman Papers, KSHS; W. E. Bush to Clifford R. Hope, April 23, 1937, Hope Papers, KSHS; William C. Washburn to Clifford R. Hope, April 6, 1937, Hope Papers, KSHS.

36. Walter Cooke to Clifford R. Hope, April 2, 1937, Hope Papers, KSHS; "Southwest Agricultural Association," SCS, General File, Drought, 1936–1937, NARG 114; Resolutions, The Southwest Agricultural Association in Session at the Court House in Boise City, Oklahoma, May 12, 1937, Hope Papers, KSHS; *Fort Worth Star-Telegram,* May 13, 1937; *Topeka Capital,* May 13, 1937.

37. *Amarillo Globe,* December 13, 1937; Rule, "Land Facts on the Southern Great Plains," pp. 13, 22; "Report of the Twentieth Conference of the Regional Advisory Committee on Land Use Practices in the Southern Great Plains Area," Amarillo, Texas, April 21–22, 1939 (mimeographed), p. 1, Soil Conservation Service Publications, NARG 114; Tannehill, *Drought: Its Causes and Effects,* p. 51.

38. Untitled manuscript, Soil Conservation Service Drought File, 1936–1937, NARG 114; Bennett, *Soil Conservation,* pp. 738, 747; Hinde, "Strip Cropping of Corn, Cotton and Beans," p. 9, NARG 114; *Amarillo Sunday News and Globe,* September 1 and November 10, 1935, April 12 and June 14, 1936.

39. Finnell, "Annual Report, 1939–1940," pp. 4, 12.

40. *Amarillo Sunday News and Globe*, December 12, 1937; "A Report of the Soil Conservation Service to the Secretary of Agriculture on Problems of the Southern Great Plains and a Conservation Program for the Region," pp. 1–6, NAL.

CHAPTER 6

1. Stewart, "Changes on Wheat Farms in Southwestern Kansas," p. 8.

2. Ibid., pp. 8–9, 23.

3. Floyd, "A History of the Dust Bowl," p. 48; "Natural and Economic Factors Which Affect Rural Rehabilitation of the North Plains of Texas (as typified by Dallam County, Texas), July, 1936," Resettlement Administration, Research Bulletin, Soil Conservation Service, General File, Drought 1936–1937, NARG 114.

4. *Amarillo Sunday News and Globe*, June 25, 1933.

5. "Memorandum for Mr. Paul Appleby," Office of the Secretary of Agriculture, August 24, 1933, from Niles A. Olsen, Chief, Bureau of Agricultural Economics, General Correspondence, NARG 114; *Amarillo Globe*, August 2, 23, 24 and September 10, 1933.

6. *Amarillo Sunday News and Globe*, January 21 and March 4, 1934; Stewart, "Changes on Wheat Farms in Southwestern Kansas," p. 9.

7. *Amarillo Sunday News and Globe*, April 29 and May 13 and 20, 1934; "The Drought of 1934: A Report of the Federal Government's Assistance to Agriculture as of July 15, 1935," Agricultural Stabilization and Soil Conservation Service, NARG 145.

8. "Oklahoma Drought," by D. P. Trent, Director, Oklahoma Extension Service, Office of the Secretary of Agriculture, General Correspondence, Drought File, 1934, NARG 16; *Amarillo Globe*, July 18, 1934, and July 1, 1935; "There's Still a Drought," *Business Week* (November 17, 1934), p. 12; *Amarillo Daily News*, April 13, 1935; *Ft. Worth Star-Telegram* (Morning), April 12, 1935; "Dust and the Nation's Breadbasket," p. 10; Beeson Museum Scrapbook, July 16, 1935, KSHS.

9. "Current Crop Prospects," July 30, 1936, Soil Conservation Service, General File, Drought 1936-1937, NARG 114; Stewart, "Changes on Wheat Farms in Southwestern Kansas," p. 8.

10. Ibid., pp. 8–9.

11. Ibid., pp. 10–11.

12. Wayne D. Rasmussen and Gladys L. Baker, *The Department of Agriculture* (New York: Praeger, 1972), pp. 23–29; Alfred H. Kelly

and Winfred A. Harbison, *The American Constitution: Its Origin and Development* (New York: W. W. Norton, 1963), pp. 744–47; Stewart, "Changes on Wheat Farms in Southwestern Kansas," p. 11.

13. Gilbert C. Fite, "Farmer Opinion and the Agricultural Adjustment Act, 1933," *Mississippi Valley Historical Review,* 48 (March 1962), pp. 668, 673; Stewart, "Changes on Wheat Farms in Southwestern Kansas," pp. 17–18.

14. "A Survey of Baca County, Colorado," pp. 34–36, NARG 83; "A Survey of Hodgeman County, Kansas," August 1, 1934, Bureau of Agricultural Economics, Rural Problem Area Reports, p. 71, NARG 83; "A Survey of Meade County, Kansas," p. 70, NARG 83; "A Survey of Cimarron County, Oklahoma," August 1934, Bureau of Agricultural Economics, Rural Problem Area Reports, p. 107, NARG 83.

15. *Amarillo Sunday News and Globe,* March 18, 1933, and January 13, March 10, and December 22, 1935; Svobida, *An Empire of Dust,* p. 50; *Elkhart* (Kansas) *Tri-State News,* February 7, 1935; *Springfield* (Colorado) *Democrat-Herald,* March 21, 1935; "A Survey of Meade County, Kansas," p. 71, NARG 83; "A Survey of Hodgeman County, Kansas," p. 71, NARG 83; "A Survey of Cimarron County, Oklahoma," p. 155, NARG 83.

16. "AAA Announces Emergency Wind Erosion Control Program for Dust Bowl Area," Agricultural Adjustment Administration, news release, April 5, 1937.

17. *Amarillo Sunday News and Globe,* December 18, 1935, and February 9 and 16, and April 2, August 2 and 10, 1936; "Survey of Current Changes in Rural Relief Population, June 1935, Greenwood County, Kansas," Rural Cooperative Research, Bureau of Agricultural Economics, Rural Relief Studies, p. 29, NARG 83; "Survey of Current Changes in the Rural Relief Population, June 1935, Seward County, Kansas," Rural Cooperative Research, Bureau of Agricultural Economics, Rural Relief Studies, p. 33, NARG 83; "Survey of Current Changes in Rural Relief Population, June 1935, Pawnee County, Kansas," Rural Cooperative Research, Bureau of Agricultural Economics, Rural Relief Studies, pp. 33–36, NARG 83; *Ft. Worth Star-Telegram* (evening), May 26, 1937; *Elkhart* (Kansas) *Tri-State News,* October 7, 1938.

18. "Aid Available to Farmers in Controlling Wind Erosion," Hope Papers, KSHS.

19. "A Survey of Roberts County, Texas," September 1934, Bureau of Agricultural Economics, Rural Problem Area Reports, p. 75, NARG 83; "A Survey of Cheyenne County, Colorado," September 28, 1934, Bureau of Agricultural Economics, Rural

Problem Area Reports, p. 40, NARG 83; "Survey of Current Changes in the Rural Relief Population, June, 1935, Hamilton County, Kansas," Rural Cooperative Research, Bureau of Agricultural Economics, Rural Relief Studies, pp. 8–9, NARG 83; "A Survey of Cimarron County, Oklahoma," p. 6, NARG 83.

20. "The Dust Bowl: Agricultural Problems and Solutions," p. 17; Kifer, *Farming Hazards in the Drought Area,* pp. 100–101.

21. "A Survey of Cimarron County, Oklahoma," pp. 5–6, 81–82, 112, NARG 83; "A Survey of Randall County, Texas," September 1934, Bureau of Agricultural Economics, Rural Problem Area Reports, p. 39, NARG 83; Floyd, "A History of the Dust Bowl," p. 73.

22. Logsdon, "The Dust Bowl," pp. 11–12; *Fifteenth Census of the United States: 1930, Population,* vol. 3, pt. 2, pp. 564–65, 569; *Sixteenth Census of the United States: 1940, Population,* vol. 2, pt. 5, pp. 862, 866–67.

23. Paul S. Taylor and Tom Vasey, "Drought Refugee and Labor Migration to California, June-December, 1935," *Monthly Labor Review,* 43 (February 1936), p. 314; Logsdon, "The Dust Bowl," pp. 12, 14.

24. *Amarillo Times,* June 25, 1940; Alsup, "A History of the Panhandle of Texas," p. 220; "The Dust Bowl: Agricultural Problems and Solutions," p. 16; *Dodge City Daily Globe,* March 11, 1935; Rule, "Crops Against the Wind on the Southern Great Plains," p. 5; Adams, "Social Behavior in a Drought-Stricken Panhandle Community," p. 38.

25. "The Dust Bowl: Agricultural Problems and Solutions," p. 16; Stewart, "Changes on Wheat Farms in Southwestern Kansas," p. 20.

26. *Springfield* (Colorado) *Democrat-Herald,* March 24, 1938; *Elkhart* (Kansas) *Tri-State News,* September 16 and December 16, 1938.

27. *Amarillo Globe,* May and June, 1939; *Elkhart* (Kansas) *Tri-State News,* June 2, 1939, and February 23, 1940.

CHAPTER 7

1. *The Beef-Cattle Program* (Washington, D.C.: U.S. Government Printing Office, 1934), pp. 1–3.

2. R. L. Wells et al., to Marvin Jones, June 8, 1933, "Regulations Relative to Loans for Feed for Farm Livestock," Farm Credit Administration, St. Louis, Missouri, June 15, 1933, Landon Papers, KSHS; *Amarillo Globe,* June 15, 1933, "Summary of 1934 Emergency Feed Loans (Drought) in the States and Counties Designated

as Drought Territory After September 1, 1935," Office of the Secretary of Agriculture, Drought File, NARG 16.

3. Governor Alfred M. Landon to F. A. Winfrey, Acting Manager, Red Cross, Washington, D.C., June 17, 1933, Landon Papers, KSHS; *Elkhart* (Kansas) *Tri-State News,* June 22, 1933.

4. "Reduced Freight Rates Account Drought Conditions," June 28, 1933, Landon Papers, KSHS; "Outline Summary of the Drought Situation," August 24, 1933, Office of the Secretary of Agriculture, General Correspondence, NARG 16; Untitled Typescript, Landon Papers, KSHS.

5. "The Drought of 1934," pp. 16, 23, 31, NARG 145.

6. "The Cattle Purchases by Agricultural Adjustment Administration, June 1934 to February 1935," A Report to G. B. Thorne, Director, Division of Livestock and Feed Grains by Harry Petrie, Chief, Cattle and Sheep Section, p. 1, Agricultural Stabilization and Soil Conservation Service, NARG, 145; John T. Schlebecker, *Cattle Raising on the Plains, 1900–1961* (Lincoln: University of Nebraska Press, 1963), p. 136; C. Roger Lambert, "The Drought Cattle Purchase, 1934–1935: Problems and Complaints," *Agricultural History,* 45 (April 1971), p. 85.

7. "The Drought Situation, June 1, 1934," Office of the Secretary of Agriculture, General Correspondence, NARG 16; "The Drought Situation, July 1, 1934," Office of the Secretary of Agriculture, General Correspondence, NARG 16.

8. C. Roger Lambert, "Texas Cattlemen and the AAA, 1933–1935," *Arizona and the West,* 14 (Summer 1972), p. 143; Lambert, "The Drought Cattle Purchase," p. 85; D. A. Fitzgerald, *Livestock Under the AAA* (Washington: The Brookings Institution, 1935), pp. 193, 196-99; "The Cattle Purchases by Agricultural Adjustment Administration," pp. 17, 19–20, NARG 145.

9. Fitzgerald, *Livestock Under the AAA,* pp. 62, 200–201; Lambert, "Texas Cattlemen and the AAA," p. 145; "The Drought of 1934–1935," Prepared in the Statistical and Historical Unit, Commodities Purchase Section, Agricultural Adjustment Administration, As of December 31, 1935, p. 51, Agricultural Stabilization and Soil Conservation Service, NARG 145.

10. "The Drought of 1934," pp. 46, 56–57, 61–64, NARG 145; "Detailed Report of the Emergency Drought Situation and Measures which have been taken or are Recommended by Federal Agencies," May 25, 1934, p. 4, Agricultural Stabilization and Soil Conservation Service, NARG 145.

11. "The Drought of 1934," p. 46; Inspector in Charge, Bureau of Animal Industry, Albuquerque, New Mexico to Chief, Bureau

of Animal Industry, Washington, D.C., September 20, 1934, AAA Commodities Purchase Section, NARG, 124.

12. "The Drought of 1934," p. 72, NARG 145.

13. Ibid., p. 76; "The Cattle Purchases by Agricultural Adjustment Administration," pp. 53–54, NARG 145.

14. "The Drought of 1934," p. 64, NARG 145.

15. Irvin M. May, Jr., "Cotton and Cattle: The FSRC and Emergency Work Relief," *Agricultural History,* 46 (July 1972), pp. 408–9.

16. "Report, Drought Cattle Operations, The State of Kansas," The Kansas Emergency Relief Committee, May 1, 1935, pp. 9, 11–12, KSHS.

17. Ibid., pp. 11–12.

18. Ibid., pp. 27, 44.

19. *Amarillo Sunday News and Globe,* January 20, 1935; *Accomplishments and Results of Government Relief, Cattle Buying Program 1934–1935,* Agricultural Stabilization and Soil Conservation Service, NARG 145; Inspector in Charge, Bureau of Animal Industry, Albuquerque, New Mexico, to Chief, Bureau of Animal Industry, September 20, 1934, AAA Commodities Purchase Section, NARG 124; *Dodge City Journal,* January 31, 1935; *Springfield* (Colorado) *Democrat-Herald,* March 21, 1935.

20. "The Drought of 1934," pp. 178–85, NARG 145; *Dodge City Journal,* January 31, 1935; *Amarillo Sunday News and Globe,* January 6, 13, and 20, 1935; "Report, Drought Cattle Operations, The State of Kansas," p. 1; Schlebecker, *Cattle Raising on the Plains,* p. 162.

21. "The Drought of 1934," pp. 47, 76–77, 102, 113–114, NARG 145; Governor Alfred M. Landon to Lawrence Westbrook Assistant Administrator, FERA, August 4, 1934, Landon Papers, KSHS; J. F. Weatherby to Senator Carl A. Hatch, September 21, 1934, Agricultural Stabilization and Soil Conservation Service, NARG 145.

22. H. Umberger to Governor Alfred M. Landon, August 27, 1934, Landon Papers, KSHS.

23. "The Drought of 1934," p. 68, NARG 145; "The Drought Situation, June 1, 1934," NARG 16; "Effects of the Drought of 1934 on Feed, Forage and Livestock," October 1934, NAL.

24. *Amarillo Sunday News and Globe,* September 30, 1934; "Selling Thistle Hay," Landon Papers, KSHS; *Dodge City Journal,* February 21 and March 4, 1935.

25. *Amarillo Sunday News and Globe,* December 9, 1934 and January 20, 1935; *Dodge City Daily Globe,* January 7, 1935.

26. "The Drought of 1934," pp. 113–14, NARG 145.

27. "Triple 'A' Facts," Landon Papers, KSHS; *Elkhart* (Kansas) *Tri-State News*, March 14, 1935; H. J. Gramlich, Special Advisor on Feed Commodities Purchase Section, Memorandum to Colonel Murphy, June 26, 1935, Surplus Marketing Administration, Records of the Livestock Disposal and Drought Program, NARG 124.

28. "The Drought Situation, April 1935," Office of the Secretary of Agriculture, General Correspondence, Drought File, NARG 16; Mike Quinn to the Hon. Phil Ferguson, April 30, 1935, AAA Subject Correspondence, Drought (July-December, 1935), NARG 145.

29. "Drought Relief Committee," Office of the Governor, August 7, 1936, Landon Papers, KSHS; Telegram, August 31, 1936, Landon Papers, KSHS; *Amarillo Globe*, February 3 and December 31, 1937; Governor Walter Huxmon to A. W. Lefeber, November 27, 1937, Huxmon Papers, KSHS.

30. Untitled news release from the Kansas Emergency Relief Committee, February 27, 1937, Landon Papers, KSHS; L. H. Hauter, Regional Director, FSA, Amarillo, Texas to W. W. Alexander, Administrator, FSA, August 9, 1937, Records of the Farmers Home Administration, Farm Security Administration, General Correspondence, 1935–1940, Appeals for Aid, NARG 96.

31. *Amarillo Globe*, August 16 and September 8 and 26, 1937.

32. "Report of the Eighteenth Conference of the Regional Advisory Committee on Land Use Practices in the Southern Great Plains Area," Stillwater, Oklahoma, October 21–22, 1938, APL; Arthur Capper Radio Address, July 17, 1936, Capper Papers, KSHS.

33. Clifford R. Hope to H. Umberger, February 23, 1937, Hope Papers, KSHS.

34. Archil F. Cyr to Clifford R. Hope, March 15, 1937, Hope Papers, KSHS; Clifford R. Hope to E. C. Summers, June 1, 1937, Hope Papers, KSHS.

35. "Report of the Fourteenth Conference of the Regional Advisory Committee on Land Use Practices in the Southern Great Plains Area," Amarillo, Texas, December 13–14, 1937.

36. *Ft. Worth Star-Telegram*, February 20, 1938; *Elkhart* (Kansas) *Tri-State News*, November 26, 1937, and January 7, February 11, and March 4, 1938.

37. Z. W. Johnson to Clifford R. Hope, January 6, 1939, Hope Papers, KSHS; Clifford R. Hope to Archil F. Cyr, January 18, 1939, Hope Papers, KSHS; Ray Jackson to Clifford R. Hope, January 13, 1939, Hope Papers, KSHS.

38. J. S. Shastid to Clifford R. Hope, January 14, 1939, Hope Papers, KSHS; *Elkhart* (Kansas) *Tri-State News,* February 9, 1940; *Ft. Worth Star-Telegram* (morning), August 9, 1940. The total land purchases were: Kansas—Morton County, 140,572 acres; Colorado—Weld County, 203, 527 acres; Otero County, 166,000 acres; Baca County, 228,374 acres; New Mexico—Harding County, 67,629 acres; Tri-state area of Cimarron County, Oklahoma, Dallam County, Texas, and Union County, New Mexico, 133,876 acres.

39. Savage, *Drought Survival of Native Grass Species,* pp. 29, 39, 44; F. E. Mollin, *If and When It Rains: The Stockman's View of the Range Question* (Denver: American National Livestock Association, 1938), pp. 41, 60; *Elkhart* (Kansas) *Tri-State News,* September 30, 1938; Lambert, "The Drought Cattle Purchase, 1934–35," p. 93.

CHAPTER 8

1. Wilmon H. Droze, *Trees, Prairies, and People* (Denton, Tex.: Texas Woman's University, 1977), pp. 50–51; E. N. Munns and J. H. Stoeckeler, "How Are the Great Plains Shelterbelts?" *Journal of Forestry,* 44 (April 1946), p. 237.

2. "Forestry for the Great Plains," U.S. Department of Agriculture, Forest Service, Lincoln, Nebraska, September 15, 1937, p. 1, Records of the Prairie States Forestry Project, NARG 114; Gilbert C. Fite, *The Farmers' Frontier* (New York: Holt, Rinehart and Winston, 1966), p. 20; Everett Dick, *Conquering the Great American Desert: Nebraska* (Lincoln: Nebraska State Historical Society, 1975), pp. 122–25.

3. Thomas R. Wessel, "Prologue to the Shelterbelt," *Journal of the West,* 6 (January 1967), p. 126.

4. Dick, *Conquering the Great American Desert,* pp. 129–31; Willis Conner Sorensen, "The Kansas National Forest, 1905–1915," *Kansas Historical Quarterly,* 35 (Winter 1969), pp. 386–95; Droze, *Trees, Prairies, and People,* pp. 41, 44, 47.

5. Ibid., pp. 62, 66, 74–75; "Forestry for the Great Plains," p. 2, NARG 114. The term "Shelterbelt Project" was chosen because it was commonly used in the Great Plains to signify a tree windbreak. A shelterbelt may consist of one or a number of tree rows usually oriented at a right angle to the prevailing winds.

6. *Amarillo Globe,* June 28 and July 22, 1934; Droze, *Trees, Prairies, and People,* p. 81; Edgar B. Nixon, *Franklin D. Roosevelt & Conservation, 1911–1945,* Vol. 2 (Washington, D.C.: Government Printing Office, 1957), p. 319; "Forestry for the Great Plains," p. 2, NARG 114.

7. Munns and Stoeckeler, "How Are the Great Plains Shelterbelts?" p. 237; "Pros and Cons of the Shelterbelt," *American Forests,* 40 (November 1934), p. 545; H. H. Chapman, "The Shelterbelt Tree Planting Project," *Journal of Forestry,* 32 (November 1934), pp. 801–2; H. H. Chapman, "The Shelterbelt Project," *Journal of Forestry,* 32 (December 1934), p. 935; Royal S. Kellog, "The Shelterbelt Scheme," *Journal of Forestry,* 32 (December 1934), p. 977; C. G. Bates, "Individual Letters Received on Shelterbelt Project," *Journal of Forestry,* 32 (December 1934), pp. 957–72; Richard Pfister, "A History and Evaluation of the Shelterbelt Project," Masters thesis, University of Kansas, 1950, pp. 20–22.

8. Thomas R. Wessel, "Roosevelt and the Great Plains Shelterbelt," *Great Plains Journal,* 8 (Spring 1969), p. 58; *Amarillo Sunday News and Globe,* October 7 and December 30, 1934; "Forestry for the Great Plains," p. 5, NARG 114.

9. Droze, *Trees, Prairies, and People,* p. 82; *Amarillo Sunday News and Globe,* July 15 and 22, 1934.

10. Wessel, "Roosevelt and the Great Plains Shelterbelt," p. 58; Nixon, *Franklin D. Roosevelt and Conservation,* Vol. 1, p. 324; *Amarillo Sunday News and Globe,* August 2, 1936.

11. Jerome Dahl, "Progress and Development of the Prairie States Forestry Project," *Journal of Forestry,* 38 (April 1940), p. 301; "A Tree Belt for the Great Plains," *American Forests,* 40 (August, 1934), p. 343; Ralphael Zon, "Shelterbelts—Futile Dream or Workable Plan?" *Science,* 81 (April 26, 1935), p. 392; *Possibilities of Shelterbelt Planting in the Plains Region* (Washington, D.C.: Government Printing Office, 1935), pp. 1–9.

12. Ibid., p. 9; Pfister, "A History and Evaluation of the Shelterbelt Project," pp. 27–28.

13. Droze, *Trees, Prairies, and People,* p. 144; *Dodge City Daily Globe,* February 5 and June 20, 1935; *Elkhart* (Kansas) *Tri-State News,* April 25, 1935; James B. Lang, "The Shelterbelt Project in the Southern Great Plains—1934–1970—A Geographical Appraisal," Master's thesis, University of Oklahoma, 1970, pp. 81–82, 98–100, 111–12.

14. Lang, "The Shelterbelt Project in the Southern Great Plains," p. 100; *Report to the Chief of the Forest Service, 1935,* pp. 7, 54.

15. Ibid., p. 7; Pfister, "A History and Evaluation of the Shelterbelt Project," pp. 32–33.

16. Wessel, "Roosevelt and the Great Plains Shelterbelt," p. 59; Droze, *Trees, Prairies, and People,* pp. 129–30; *Report of the Chief of the Forest Service, 1936,* p. 42.

17. Ibid.; Glen R. Durrell, "Social and Economic Effects of the Great Plains Shelterbelt in Terms of Social and Human Betterment," *Journal of Forestry,* 37 (February 1939), p. 144.

18. Droze, *Trees, Prairies, and People,* p. 195.

19. Lang, "The Shelterbelt Project in the Southern Great Plains," p. 112; *Report of the Chief of the Forest Service, 1936,* pp. 42–43; Droze, *Trees, Prairies, and People,* p. 229 ff.

20. Nixon, *Roosevelt and Conservation,* Vol. 2, pp. 335, 446, 527; Droze, *Trees, Prairies, and People,* p. 209.

21. *Report of the Chief of the Forest Service, 1937,* pp. 50–51; Droze, *Trees, Prairies, and People,* p. 229 ff.

22. Droze, *Trees, Prairies, and People,* pp. 210–13; Wessel, "Roosevelt and the Great Plains Shelterbelt," p. 63; Dahl, "Progress and Development of the Prairie States Forestry Project," p. 301.

23. "Progress Report of the Work of the Forest Service in Kansas, July 1, 1937, to June 30, 1938," pp. 2–3, Hope Papers, KSHS; "Progress Report of the Work of the Forest Service in Kansas, July 1, 1938, to December 31, 1938," p. 1, Hope Papers, KSHS; Verna Carney Alden, "A History of the Shelterbelt Project in Kansas," Master's thesis, Kansas State College of Agriculture and Applied Science, 1949, pp. 28–29; T. Russell Reitz, "A Traveler Sees the Shelterbelts," *Progress in Kansas,* 7 (December 1940), p. 12; Pfister, "A History and Evaluation of the Shelterbelt Project pp. 51–52.

24. "Progress Report of the Work of the Forest Service in Kansas, July 1, 1937, to June 30, 1938," p. 2, Hope Papers, KSHS; *Elkhart* (Kansas) *Tri-State News,* April 29, 1938; Lang, "The Shelterbelt Project in the Southern Great Plains," pp. 81–82; "Shelterbelt Planting for 1938," *Journal of Forestry,* 36 (June 1938), p. 581.

25. Edwin R. Henson, Coordinator, Southern Great Plains Prairie States Forestry Project to Milton S. Eisenhower, Land Use Coordinator, Department of Agriculture, Washington, D.C., December 15, 1939, Records of the Prairie States Forestry Project, NARG 114; *Dodge City Daily Globe,* October 10, 1935.

26. Wessel, "Roosevelt and the Great Plains Shelterbelt," p. 68; Droze, *Trees, Prairies, and People,* pp. 217–19, 222.

27. *Report of the Chief of the Forest Service, 1942,* p. 22; Pfister, "A History and Evaluation of the Shelterbelt Project," p. 67; Munns and Stoeckeler, "How Are the Great Plains Shelterbelts?" pp. 241–42; Henson to Eisenhower, December 15, 1939, NARG 114.

28. Droze, *Trees, Prairies, and People,* pp. 226–27, 235, 242; Munns and Stoeckeler, "How Are the Great Plains Shelterbelts?" pp. 249–50; Pfister, "A History and Evaluation of the Shelterbelt Project," pp. 69–70.

29. Lang, "The Shelterbelt Project in the Southern Great Plains," pp. 120–23; Elmer W. Shaw, "A New Look at the Shelterbelts," *American Forests,* 63 (December 1957), p. 48; Henson to Eisen-

hower, December 15, 1939, NARG 114; Munns and Stoeckeler, "How Are the Great Plains Shelterbelts?" pp. 239, 253; "Wallace Praises Prairie Tree Planting," *American Forests*, 45 (July 1939), p. 377; *Possibilities of Shelterbelt Planting in the Plains Region*, p. 7; Droze, *Trees, Prairies, and People*, p. 240.

30. J. H. Stoeckeler, "Shelterbelt Planting Reduces Wind Erosion Damages in Western Oklahoma," *Journal of the American Society of Agronomy*, 30 (November 1938), p. 931; "Forestry for the Great Plains," pp. 2–3, NARG 114; Zon, "Shelterbelts—Futile Dream or Workable Plan?" p. 394.

CHAPTER 9

1. "A Report of the Soil Conservation Service to the Secretary of Agriculture on Problems of the Southern Great Plains and a Conservation Program for the Region," p. iv, NAL; H. H. Finnell, "The Plowup of the Western Grasslands and the Resultant Effect upon Great Plains Agriculture," *Southwestern Social Science Quarterly*, 32 (September 1951), p. 96; "Facts About Wind Erosion and Dust Storms on the Great Plains," U.S Department of Agriculture, Leaflet No. 394, 1955, p. 4; Tom Dale, "When Drought Returns to the Great Plains," U.S. Department of Agriculture, Farmers' Bulletin No. 1982, 1947, p. 3; Gilbert Hill, "The Plains Will Blow Again," *Science Digest*, 17 (June 1945), p. 89.

2. "A Report of the Soil Conservation Service to the Secretary of Agriculture on Problems of the Southern Great Plains and a Conservation Program for the Region," p. 10, NAL; *Kansas City Times*, December 23, 1948; *Christian Science Monitor*, magazine section, September 11, 1948; Dwight Weinland to Governor Frank Carlson, June 6, 1949, Kansas State Historical Society Library; *Topeka Daily Capital*, April 7, 1950.

3. *Elkhart* (Kansas) *Tri-State News*, January 13 and 20, February 3, March 10, and April 7, 1950; *Kansas City Times*, March 7 and 31, 1950; *Springfield* (Colorado) *Plainsman-Herald*, March 9, 1950; *Dodge City Daily Globe*, April 8, 10, 11 and July 3, 1950; *Wichita Morning Eagle*, April 11, 1950.

4. "Facts About Wind Erosion and Dust Storms," p. 4; "A Report of the Soil Conservation Service to the Secretary of Agriculture on Problems of the Southern Great Plains and a Conservation Program for the Region," p. 2, NAL; *Springfield* (Colorado) *Plainsman-Herald*, January 3, 1952; *Topeka Journal*, February 19 and April 15, 1953; *Topeka Capital*, March 22, 1953.

5. Conference Held with Clifford R. Hope, Congressman, at the Warren Hotel in Garden City, Kansas, 1:00 o'clock P.M., February

24, 1954, Hope Papers, KSHS; *Lamar* (Colorado) *Daily News,*
March 1 and 6, 1954.

6. *Topeka Capital,* March 7, 1954; *Kansas City Star,* March 16 and
18, 1954; *Fact Sheet on Drought and Storm Damage,* March 2, 1954,
Hope Papers, KSHS. The seven counties referred to in the March
report are Baca, Bent, Cheyenne, Lincoln, Kiowa, Prowers, and
Cowley. The dollar estimate of the wheat loss is based on a price of
$2.00 per bushel.

7. "A Report of the Soil Conservation Service to the Secretary of
Agriculture on Problems of the Southern Great Plains and a Con-
servation Program for the Region," pp. 2–3, NAL; "Facts About
Wind Erosion and Dust Storms," p. 4; *Elkhart* (Kansas) *Tri-State
News,* February 26 and June 18, 1954; *Dodge City Daily Globe,* Feb-
ruary 19 and 20, 1954; W. S. Chepil and N. P. Woodruff, "How to
Reduce Dust Storms," Kansas Agricultural Experiment Station,
Manhattan, Circular 318, 1955, p. 5; C. B. Palmer, "Out Where the
Dust Begins," *New York Times Magazine,* April 11, 1954, pp. 40,
42.

8. *Topeka Capital,* March 11, 1955; "Wind Erosion Conditions—
Great Plains Summary of Local Estimates as of March 1, 1955,"
Hope Papers, KSHS; "Wind Erosion Conditions—Great Plains
Summary of Local Estimates as of December 1, 1955," Hope
Papers, KSHS; Leonard Lindholn to Clifford R. Hope, December
15, 1955, Hope Papers, KSHS: *New York Times,* April 1, 1955.

9. *Topeka Capital,* March 28 and 29, 1956.

10. *Elkhart* (Kansas) *Tri-State News,* March 30, 1956; *Dodge City
Daily Globe,* April 3, 1956; *Topeka Capital,* October 25, 1956 and
March 12, 1957.

11. *Lamar* (Colorado) *Daily News,* January 31 and February 8,
1952, and March 22 and 28, 1954; *Elkhart* (Kansas) *Tri-State News,*
January 1 and March 12, 1954; *Dodge City Daily Globe,* March 1,
1954. As of January 1, 1950, Soil Conservation Districts protected
250,537,000 acres in the Dust Bowl, see *Topeka Capital,* April 7,
1950.

12. Clifford R. Hope to Governor Edward F. Arn, March 25,
1954, Hope Papers, KSHS; *Kansas City Times,* March 12 and May
14, 1954; Clifford R. Hope to George L. Lizer, May 14, 1954,
Hope Papers, KSHS; *Garden City* (Kansas) *Times,* May 14, 1954;
Ulysses (Kansas) *News,* May 20, 1954.

13. *Elkhart* (Kansas) *Tri-State News,* June 18, 1954; Clifford R.
Hope to Governor Edward F. Arn, March 25, 1954, Hope Papers,
KSHS; Clifford R. Hope to Howard Hooker, March 5, 1955, Hope
Papers, KSHS; *Topeka Journal,* April 7, 1955; "USDA Reports

Additional Wind Damage in Southern Great Plains," April 25, 1955, Hope Papers, KSHS.

14. "Wind Erosion Conditions—Great Plains Summary of Local Estimates, June 1, 1955," Hope Papers, KSHS; "Wind Erosion Conditions—Great Plains Summary of Local Estimates as of March 1, 1955." Of the 17,116,000 acres estimated to be in a condition to blow, 13,336,000 acres were in crop land and 3,383,000 acres were in range land. "Wind Erosion Conditions—Great Plains Summary of Local Estimates as of April 1, 1955," Hope Papers, KSHS.

15. "USDA Allocates $4,275,000 from Disaster Fund for Wind Erosion Work in Six States," April 7, 1955, Hope Papers, KSHS; "Wind Erosion Conditions—Great Plains Summary of Local Estimates as of May 1, 1955," Hope Papers, KSHS; "Wind Erosion Conditions—Great Plains, June 1, 1955," Hope Papers, KSHS; "Wind Erosion Conditions—Great Plains Summary of Local Estimates as of December 1, 1955," Hope Papers, KSHS.

16. B. W. Allred and W. M. Nixon, "Grass for Conservation in the Southern Great Plains," U.S. Department of Agriculture, Farmers' Bulletin No. 2093, 1955, p. 1.

17. Allred and Nixon, "Grass for Conservation in the Southern Great Plains," pp. 3, 18, 19; H. G. Reynolds, "Reseeding Southwestern Range Lands with Crested Wheatgrass," U. S. Department of Agriculture, Farmers' Bulletin No. 2056, 1953.

18. U.S. Congress, House of Rep., Committee on Agriculture, Hearings of the Severe Drought Situation in the Southwestern Area of the United States, 83rd. Cong., 1st sess., 1953; "Emergency Drought Feed Program Summary for Week Ending July 17, 1953," Hope Papers, KSHS.

19. Schlebecker, *Cattle Raising on the Plains, 1900–1961,* pp. 207–8; Hearings on the Severe Drought Situation in the Southwestern Area of the United States.

20. Hearings on the Severe Drought Situation in the Southwestern Area of the United States.

21. "Mixed Feeds Made Available at Reduced Prices in Drought Disaster Area," USDA, Production Marketing Administration, July 24, 1953, Hope Papers, KSHS; *Wichita Eagle,* August 14, 1953.

22. "Drought Emergency Hay Program," memorandum from Wendell Becraft to Chairmen, PMA Committees in Drought Area, USDA, Production Marketing Administration, Manhattan, Kansas, October 29, 1953, Hope Papers, KSHS.

23. "1954 U.S. Department of Agriculture Drought Program," August, 1954, pp. 4–5, Hope Papers, KSHS.

24. Ibid., p. 5; "The 1954 Emergency Feed Program," October 1, 1954, Hope Papers, KSHS; Wendell Becraft to Paul P. Pancy, June 6, 1955, Hope Papers, KSHS; "Designation of Emergency Loan Area in Great Plains Extended," September 16, 1955, Hope Papers, KSHS. Later, the FHA allowed a reduction of $1.00 per hundred weight of feed grain.

25. "Feb. 15 Set as Deadline to Apply for Feed Assistance Under Drought Programs," Hope Papers, KSHS.

26. "1954 U.S. Department of Agriculture Drought Program," pp. 1–2, Hope Papers, KSHS.

27. Ibid., pp. 2–3, 6; *Kansas City Star,* May 1, 1955. In spring, 1955, the FHA raised the interest rate from 3 percent to 5 percent. Some farmers were abusing the program by taking low interest loans from the FHA while avoiding other sources. Since the FHA was to be a noncompetitive lending agency, the interest was raised to discourage those farmers from borrowing who did not have a legitimate reason to do so.

28. "1954 U.S. Department of Agriculture Drought Program," p. 3, Hope Papers, KSHS.

29. Earl W. Chapman to Clifford R. Hope, May 28, 1957, Hope Papers, KSHS.

30. *Topeka Journal,* April 26, 1955; Andrew B. Erhart, "How Far for Irrigation in Kansas?" Thirty-Eighth Biennial Report of the Kansas State Board of Agriculture (1951–1952), p. 25; *Amarillo Sunday News and Globe,* October 21, 1934, and April 14, May 17, July 12, 14, and 26, August 2, September 22, and October 25, 1935; Donald E. Green, *Land of the Underground Rain* (Austin: University of Texas Press, 1973), pp. 124–25; George S. Knapp, "Report of the Division of Water Resources," Thirty-First Biennial Report of the Kansas State Board of Agriculture (1937–1938), p. 148; Richard Pfister, *Water Resources and Irrigation,* Economic Development in Southwestern Kansas, part iv (Lawrence: University of Kansas School of Business, Bureau of Business Research, 1955). In Colorado, the irrigated acreage actually decreased between 1930 and 1950. Nevertheless, in 1950, Colorado still ranked third in total irrigated acres behind California and Texas. See *A Hundred Years of Irrigation in Colorado* (Denver: Colorado Water Conservation Board and Colorado Agricultural and Mechanical College, 1952), p. 72.

31. Russell L. Herpich, "Kansas Irrigation Potential," Forty-Second Annual Report of the Kansas State Board of Agriculture (July 1958 to June 1959), pp. 50–51; Russell L. Herpich and R. D. McKinney, "Irrigation Farming for Profit," Manhattan Agricultural

Experiment Station, Circular No. 372, 1959, pp. 3–4; Kansas Water Resources Fact Finding and Research Committee, "Water in Kansas: A Report to the 1955 Kansas State Legislature," July 26, 1954, p. 53; Ivan D. Wood, "Irrigation in the Middle West," *Agricultural Engineering,* 38 (June 1957), p. 419; Pfister, *Water Resources and Irrigation,* pp. 82, 85; *Topeka Capital,* June 14, 1953; John J. Penney, John A. Anderson, and Donald F. Kostechi, "Little Arkansas River Basin," Kansas Water Resources Board, State Water Plan Studies, Part C, 1975, pp. 30–31.

    32. Green, *Land of the Underground Rain.*

    33. Margaret Bourke-White, "Dust Plague Upon the Land," *Life,* 36 (May 3, 1957), p. 35; *Dodge City Daily Globe,* March 30, 1956; *Elkhart* (Kansas) *Tri-State News,* March 12, 1954; *Topeka Capital,* March 7, 1954; *Kansas City Times,* May 31, 1950; *Topeka Journal,* April 26, 1955.

    34. Hewes, *Suitcase Farming Frontier,* pp. 112, 119, 131, 173.

    35. R. Douglas Hurt, "Dust Bowl," *The American West,* July/August 1977. Copyright ©1977 by the American West Publishing Company, Cupertino, Calif. Reprinted by permission of the publisher.

EPILOGUE

    1. *Dallas Morning News,* March 7, 1965.

    2. *Washington* (D.C.) *Post,* March 26, 1976; *National Observer,* March 6, 1976; "Will It Be a Dust Bowl All Over Again?" *U.S. News & World Report,* (April 26, 1976), p. 69; Michael Graznak, "Too Early for Hysteria in Drought-Hit Wheat Areas," *Farmland News,* (March 15, 1976), p. 12.

    3. *Topeka Capital,* February 24 and 25, 1977; *The Marketeer,* 9 (April 1977), p. 5; *Hays* (Kansas) *Daily News,* March 9, 1977; *Washington* (D.C.) *Post,* March 26, 1977; Glenn Lorang, "Wanted: Water for the Plains and West," *Farm Journal,* 101 (March 1977), p. 32B, 33A; "Remember the 30's, 50's," *The Marketeer,* 9 (April 1977), p. 12.

    4. *Hays* (Kansas) *Daily News,* June 23, 1977.

    5. Personal observations of the author; *Lubbock Avalanche-Journal,* December 16, 1977.

    6. R. Douglas Hurt, "Dust!" *American Heritage,* 28 (August 1977), p. 35.

Bibliography

MANUSCRIPTS AND ARCHIVAL MATERIALS

Amarillo, Texas. Jeanne Claytor Collection.
Amarillo, Texas. Amarillo Public Library. John L. McCarty Collection.
Lubbock, Texas. Southwest Collection, Texas Tech University. W. G. Deloach Papers, Diary 6.
Norman, Oklahoma. Western History Collections, University of Oklahoma. W. S. Campbell Collection.
Topeka, Kansas. Kansas State Historical Society. Arthur Capper Papers.
Topeka, Kansas. Kansas State Historical Society. Clifford R. Hope Papers.
Topeka, Kansas. Kansas State Historical Society. Walter Huxmon Papers.
Topeka, Kansas. Kansas State Historical Society. Alfred M. Landon Papers.
Topeka, Kansas. Kansas State Historical Society. Report, Drought Cattle Operations, The State of Kansas, by the Emergency Relief Committee, Topeka, Kansas, May 1, 1935.
Washington, D.C. National Archives. Records of the Agricultural Adjustment Administration.
Washington, D.C. National Archives. Records of the Agricultural Stabilization and Soil Conservation Service.
Washington, D.C. National Archives. Records of the Commodities Purchase Section, Bureau of Animal Industry, Agricultural Adjustment Administration.
Washington, D.C. National Archives. Records of the Farm Security Administration.

Washington, D.C. National Archives. Records of the Farmers'
    Home Administration.
Washington, D.C. National Archives. Records of the Federal Emer-
    gency Relief Administration, Rural Problem Area Reports.
Washington, D.C. National Archives. Records of the Prairie States
    Forestry Project.
Washington, D.C. National Archives. Records of the Office of the
    Secretary of Agriculture.
Washington, D.C. National Archives. Records of the Soil Conser-
    vation Service.
Washington, D.C. National Archives. Records of the Surplus Mar-
    keting Administration.

PRINTED DOCUMENTS

Allred, B. W., and Nixon, W. M. "Grass for Conservation in the
    Southern Great Plains." U.S. Department of Agriculture,
    Farmers' Bulletin No. 2093, 1955.
Anderson, John A., and Kostechi, Donald F. "Little Arkansas River
    Basin." Kansas Water Resources Board, State Water Plan
    Studies, Part C, 1975.
Archer, L. C. "Curbing the Wind." Twenty-Ninth Biennial Report
    of the Kansas State Board of Agriculture (1933–1934), pp.
    67–71.
_____ . "The Fort Hays Damming Attachment for Listers." Thir-
    tieth Biennial Report of the Kansas State Board of Agriculture
    (1935–1936), pp. 78–82.
Beck, Paul G., and Forster, M. C. "Six Rural Problem Areas, Relief-
    Resources-Rehabilitation." Federal Emergency Relief Admin-
    istration, Research Monograph 1, Washington, D.C., 1935.
The Beef-Cattle Program. Washington, D.C.: Government Printing
    Office, 1934.
Bell, Earl H. Culture of a Contemporary Rural Community: Sublette,
    Kansas. Rural Life Studies: 2, U.S. Department of Agriculture,
    Bureau of Agricultural Economics, September 1942.
Bennett, H. H. Handbook of Soil and Water Conservation Practices for
    the Wind Erosion Area. U.S. Department of Agriculture, Wash-
    ington, D.C., 1936.
Brown, Earl G. "Dust Storms and Their Possible Effect on Health."
    Public Health Reports, 50 (October 4, 1935), pp. 1369–84.
Chepil, W. S. and Woodruff, N. P. "How to Reduce Dust Storms."
    Kansas Agricultural Experiment Station Circular 318, 1955.

Chilcott, E. F. "Preventing Soil Blowing on the Southern Great Plains." U.S. Department of Agriculture, Farmers' Bulletin No. 1771, 1937.

Dale, Tom. "When Drought Returns to the Great Plains." U.S. Department of Agriculture, Farmers' Bulletin No. 1982, 1947.

Erhart, Andrew B. "How Far for Irrigation in Kansas?" Thirty-Eighth Biennial Report of the Kansas State Board of Agriculture (1951–1952), pp. 25–36.

"Facts About Wind Erosion and Dust Storms on the Great Plains." U.S. Department of Agriculture, Leaflet No. 394, 1955.

Flora, Snowden D. "Climate of Kansas." Report of the Kansas State Board of Agriculture, (June 1948), pp. 1–320.

Finnell, H. H. "Dust Storms Come from Poorer Lands." U.S. Department of Agriculture, Leaflet No. 260, 1949.

Free, E. E., and Westgate, J. M. "The Control of Blowing Soils." U. S. Department of Agriculture, Farmers' Bulletin No. 421, 1910.

Grimes, W. E. "Marginal and Submarginal Lands in Kansas." Twenty-Ninth Biennial Report of the Kansas State Board of Agriculture (1933–1934), pp. 60–67.

Herpich, Russell L. "Kansas Irrigation Potential." Forty-Second Annual Report of the Kansas State Board of Agriculture (July 1958–June 1959), pp. 50–52.

————, and McKinney, R. D. "Irrigation Farming for Profit." Kansas Agricultural Experiment Station, Circular No. 372, 1959.

"How to Control Wind Erosion." Agricultural Information Bulletin No. 354, 1972.

Hoyt, John C. *Droughts of 1930–34.* Washington, D.C.: Government Printing Office, 1936.

————. "Droughts of 1936 with Discussion on the Significance of Drought in Relation to Climate." Washington, D.C.: Government Printing Office, 1938.

Joel, Arthur H. "Soil Conservation Reconnaissance Survey of the Southern Great Plains Area." U.S. Department of Agriculture, Technical Bulletin No. 556, 1937.

Kansas. *Session Laws, Ch. 189, 1937.*

Kansas ex rel. Perkins v. *Hardwick et al.,* 144 Kan. 3.

Kansas Legislative Council, Resources Department. "Soil Drifting." Publication No. 43, November 12, 1936.

Kifer, R. S., and Stuart, H. L. *Farming Hazards in the Drought Area.* WPA, Division of Social Research, Research Monograph 16, Washington, D.C., 1938.

Kimmel, Roy I. "A Long View of the Wind Erosion Problem." Quarterly Report of the Kansas State Board of Agriculture (March 1938), pp. 84–94.

Knapp, George S. "Report of the Division of Water Resources." Thirty-First Biennial Report of the Kansas State Board of Agriculture (1937–1938), pp. 148–49.

McDonald, Angus. "Erosion and Its Control in Oklahoma Territory." U.S. Department of Agriculture, Miscellaneous Publication No. 301, 1938.

Possibilities of Shelterbelt Planting in the Great Plains Region. Washington, D.C.: Government Printing Office, 1935.

Reynolds, H. G., and Springfield, H. W. "Reseeding Southwestern Range Lands with Crested Wheatgrass." U.S. Department of Agriculture, Farmers' Bulletin No. 2056, 1953.

Rule, Glenn K. "Crops Against the Wind on the Southern Great Plains." U.S. Department of Agriculture, Farmers' Bulletin No. 1833, 1939.

——————. "Land Facts on the Southern Great Plains." U.S. Department of Agriculture, Miscellaneous Publication No. 334, 1939.

Savage, D. A. "Drought Survival of Native Grass Species in the Central and Southern Great Plains." U.S. Department of Agriculture, Technical Bulletin No. 549, 1937.

"Soil Conservation Districts for Erosion Control." U.S. Department of Agriculture, Miscellaneous Publication No. 293, 1937.

The Beef-Cattle Program. Washington, D.C.: Government Printing Office, 1934.

Throckmorton, R. I. "A Soil Conservation Program for Kansas." Thirty-First Biennial Report of the Kansas State Board of Agriculture (1937–1938), pp. 39–53.

U.S. Congress, House of Rep., Committee on Agriculture, Drought Program Hearing, 83rd Cong., 2d sess., 1954.

U.S. Congress, House of Rep., Committee on Agriculture, Hearings of the Severe Drought Situation in the Southwestern Area of the United States, 83rd Cong., 1st sess., 1953.

U.S. Department of Commerce, Bureau of the Census, *Fifteenth Census of the United States, 1930:* Population, Vol. 3.

——————. *Sixteenth Census of the United States,* 1940: Characteristics of the Population, Vol. 2.

U.S. Department of Agriculture. *Yearbook of Agriculture, 1935.* Washington, D.C.: Government Printing Office, 1936.

U.S. Geological Survey, "Hydrology," *Twenty-Second Annual Report,* pt. 4, Washington, D.C., 1902.

U.S. Natural Resources Board. *Soil Erosion a Critical Problem in American Agriculture,* Part 5. Washington, D.C.: Government Printing Office, 1935.

Whitfield, Charles J. "Sand Dune Reclamation in the Southern Great Plains." U.S. Department of Agriculture, Farmers' Bulletin No. 1825, 1939.

Woodruff, N. P., Chepil, W. S., and Lynch, R. D. "Emergency Chiseling to Control Wind Erosion." Kansas Agricultural Experiment Station Technical Bulletin No. 90, 1957.

NEWSPAPERS, SCRAPBOOKS, AND AGRICULTURAL PERIODICALS

Abilene (Kansas) *Daily Reflector*
Abilene (Kansas) *Gazette*
Amarillo Daily News
Amarillo Globe
Amarillo Times
Beeson Museum Dust Bowl Scrapbook (Kansas State Historical Society)
Colorado Springs Gazette
Christian Science Monitor
Dalhart Texan
Dallas Morning News
Dawson Scrapbook (Colorado Historical Society)
Denver Post
Dodge City Daily Globe
Dodge City Journal
Dust Bowl Scrapbook (Kansas State Historical Society)
Elkhart (Kansas) *Tri-State News*
Ellis (Kansas) *County Star*
Farmland News
Ft. Worth Star-Telegram
Garden City (Kansas) *Daily Telegram*
Hays (Kansas) *Daily News*
Hoisington (Kansas) *Dispatch*
Hutchinson (Kansas) *Herald*
Kansas City Daily Drover's Telegram
Kansas City Star
Kansas City Times
Kansas Farmer
Lamar (Colorado) *Daily News*
Lamar (Colorado) *Register*
Liberal (Kansas) *News*

Lubbock Avalanche-Journal
Muleshoe (Texas) *Journal*
National Observer
New York Times
Pueblo (Colorado) *Indicator*
Rocky Mountain News
Salina (Kansas) *Journal*
Sandstorm Tribute Scrapbook (Claytor Manuscript Collection)
Southwest (Liberal, Kansas) *Tribune*
Springfield (Colorado) *Democrat-Herald*
Springfield (Colorado) *Plainsman-Herald*
Sterling (Colorado) *Advocate*
Thomas County Clippings (Kansas State Historical Society)
Topeka Daily Capital
Topeka Journal
Ulysses (Kansas) *News*
Washington (D.C.) *Evening Star*
Washington (D.C.) *Post*
Wichita Eagle
Winfield (Kansas) *Daily Courier*

BOOKS

Bennett, Hugh Hammond. *Soil Conservation.* New York: McGraw-
   Hill, 1939.
Bonnifield, Paul. *The Dust Bowl: Men, Dirt, and Depression.* Albu-
   querque: University of New Mexico Press, 1979.
Burges, Austin E. *Soil Erosion and Control.* Atlanta: Turner E. Smith,
   1936.
Dick, Everett. *Conquering the Great American Desert: Nebraska.* Lin-
   coln: Nebraska State Historical Society, 1975.
Droze, Wilmon H. *Trees, Prairies, and People.* Denton, Tex.: Texas
   Woman's University, 1977.
Fite, Gilbert C. *The Farmers' Last Frontier, 1865–1900.* New York:
   Holt, Rinehart and Winston, 1966.
Fitzgerald, D. A. *Livestock Under the AAA.* Washington, D.C.: The
   Brookings Institution, 1935.
French, Warren. *A Companion to the Grapes of Wrath.* New York:
   Viking Press, 1963.
Goodrich, Carter, et al. *Migration and Economic Opportunity.* Phila-
   delphia: University of Pennsylvania Press, 1936.
Green, Donald E. *Land of the Underground Rain.* Austin: University
   of Texas Press, 1973.

Hewes, Leslie. *The Suitcase Farming Frontier.* Lincoln: University of Nebraska Press, 1973.

A Hundred Years of Irrigation in Colorado. Denver: Colorado Water Conservation Board and Colorado Mechanical College, 1952.

Johnson, Vance. *Heaven's Tableland.* New York: Farrar, Straus, 1947.

Kelly, Alfred H. and Haribson, Winfred A. *The American Constitution: Its Origins and Development.* New York: W. W. Norton, 1963.

Kraenzel, Carl Frederick. *The Great Plains in Transition.* Norman: University of Oklahoma Press, 1969.

Lord, Russell. *The Care of the Earth: A History of Husbandry.* New York: Mentor Books, 1963.

Malin, James C. *The Grasslands of North America: Prolegomena to Its History with Addenda.* Lawrence, Kan.: N.P., 1961.

————. *Winter Wheat in the Golden Belt of Kansas.* Lawrence: University of Kansas Press, 1944.

Mays, William E. *Sublette Revisited: Stability and Change in a Rural Kansas Community After a Quarter Century.* New York: Florham Park Press, 1968.

Nixon, Edgar B. *Franklin D. Roosevelt & Conservation 1911–1945.* Washington, D.C.: Government Printing Office, 1957.

Pfister, Richard. *Water Resources and Irrigation.* Economic Development of Southwestern Kansas, Part iv. Lawrence: University of Kansas School of Business, Bureau of Business Research, 1955.

Rister, Carl Coke. *No Man's Land.* Norman: University of Oklahoma Press, 1948.

Schlebecker, John T. *Cattle Raising on the Great Plains, 1900–1961.* Lincoln: University of Nebraska Press, 1963.

Sears, Paul B. *Deserts on the March.* Norman: University of Oklahoma Press, 1935.

Smith, H. T. U. *Geological Studies in Southwestern Kansas.* Lawrence: University of Kansas Press, 1940.

Stein, Walter J. *California and the Dust Bowl Migration.* Westport, Conn.: Greenwood Press, 1973.

Svobida, Lawrence. *An Empire of Dust.* Caldwell, Idaho: Caxton Printers, 1940.

Tannehill, Ivan Ray. *Drought: Its Causes and Effects.* Princeton: Princeton University Press, 1947.

Vestal, Stanley. *Short Grass Country.* New York: Duell, Sloan & Pearce, 1941.

Weaver, J. E. and Albertson, F. W. *Grasslands of the Great Plains.* Lincoln: Johnsen Publishing Co., n.d.

Webb, Walter Prescott. *The Great Plains.* New York: Grosset & Dunlap, 1931.

Wilson, Robert R. and Sears, Ethel M. *History of Grant County, Kansas.* Wichita: Wichita Eagle Press, 1950.

Wolfert, Ira. *An Empire of Genius.* New York: Simon and Schuster, 1960.

Worster, Donald. *Dust Bowl: The Southern Plains in the 1930s.* New York: Oxford University Press, 1979.

ARTICLES

"A Tree Belt for the Prairies." *American Forests,* 40 (August 1934), p. 343.

"Agriculture: 500,000,000 Tons of Dust Cover Kansas and Points East." *Newsweek* (March 30, 1935), pp. 5–6.

Barnes, Lela. "Journal of Isaac McCoy for the Exploring Expedition of 1830." *Kansas Historical Quarterly,* 5 (November 1936), pp. 339–77.

Barton, Thomas F. "The Great Plains Tree Planting Project." *Journal of Geography,* 35 (April 1936), pp. 125–35.

Bates, C. G. "Individual Letters Received on Shelterbelt Project." *Journal of Forestry,* 32 (December 1934), pp. 957–72.

————. "The Plains Shelterbelt Project." *Journal of Forestry,* 32 (December 1934), pp. 978–91.

Beddow, James B. "Depression and New Deal: Letters from the Plains." *Kansas Historical Quarterly,* 43 (Summer 1977), pp. 140–53.

Bennett, Hugh Hammond. "Facing the Erosion Problem." *Science,* 81 (April 5, 1935), pp. 321–26.

Blackmar, Fran W. "The History of the Desert." *Transactions of the Kansas State Historical Society,* 9 (1905–1906), pp. 101–14.

Bonnifield, Paul. "The Oklahoma Panhandles' Agriculture to 1930," *Red River Valley Historical Review,* 3 (Winter 1978), pp. 61–76.

Bourke-White, Margaret. "Dust Changes America," *Nation,* 140 (May 22, 1935), pp. 597–98.

————. "Dust Plague upon the Land," *Life,* 36 (May 3 1954), pp. 34–38, 41.

Bourne, Patricia M., and Sagerser, A. Bower. "Background Notes on the Bourne Lister Cultivator." *Kansas Historical Quarterly,* 20 (August 1952), pp. 183–86.

Burrill, Meredith F. "Geography and the Relief Problem in Texas and Oklahoma." *Southwestern Social Science Quarterly,* 17 (December 1936), pp. 294–302.

Butler, Ovid. "The Prairie Shelterbelt." *American Forests,* 40 (September 1934), pp. 395–98, 431.

Call, L. E. "Conditions in Western Kansas." *The Land Today and Tomorrow,* 2 (April 1935), pp. 8–11.

Carlson, Avis D. "Dust." *New Republic,* 82 (May 1, 1935), pp. 332–33.

————. "Dust Blowing." *Harper's,* 171 (July 1935), pp. 149–58.

Chapman, H. H. "Digest of Opinions Received on the Shelterbelt Project." *Journal of Forestry,* 32 (December 1934), pp. 952–57.

————. "The Shelterbelt Tree Planting Project." *Journal of Forestry,* 32 (November 1934), pp. 801–3.

Chase, Stuart. "Disaster Rides the Plains." *American Magazine,* 124 (September 1937), pp. 46–47, 66, 68, 70.

Country Gentleman, 64 (March 16, 1899), p. 207.

Dahl, Jerome. "Progress and Development of the Prairie States Forestry Project." *Journal of Forestry,* 38 (April 1940), pp. 301–6.

Davenport, Walter. "Land Where Our Children Die." *Collier's,* (September 18, 1937), pp. 12–18.

Davis, Chester C. "Lost Acres." *American Magazine,* 121 (February 1936), pp. 63, 127–29.

Dawber, Mark A. "Churches in the Dust Bowl." *Missionary Review of the World,* 62 (September 1939), pp. 394–97.

"Deep Well Irrigation in Dust Bowl." *Business Week,* (September 11, 1937), pp. 40, 42.

"Documented Dust." *Time* (May 25, 1936), pp. 47–48.

Doerr, Arthur H. "Dry Conditions in Oklahoma in the 1930's and 1950's as Delimited by the Original Thornthwaite Climatic Classification." *Great Plains Journal,* 2 (Spring 1963), pp. 67–77.

"Drought Effects Deepen." *Business Week* (August 11, 1934), pp. 5–6.

"Drought Strikes the Great Plains." *Business Week* (November 18, 1939), pp. 16, 18.

Drummond, W. I. "Dust Bowl." *Review of Reviews,* 93 (June 1936), pp. 37–40.

Dudley, F. L. "Wind Erosion in the Great Plains." *The Land Today and Tomorrow,* 2 (April 1935), pp. 5–8.

Dunbar, Robert G. "Agricultural Adjustment in Eastern Colorado in the Eighteen Nineties." *Agricultural History,* 18 (January 1944), pp. 41–52.

Duncan, Kunigunde. "Reclaiming the Dust Bowl." *Nation,* 149 (September 9, 1939), pp. 269–71.

Durrell, Glen R. "Social and Economic Effects of the Great Plains Shelterbelt in Terms of Social and Human Betterment." *Journal of Forestry,* 37 (February 1939), pp. 144–47.

"Dust Bowl into Grazing Land." *Literary Digest,* 121 (March 7, 1936), p. 9.

"Dust and Politics." *Business Week* (April 20, 1935), pp. 15–16.

" Dust and the Nation's Bread Basket." *Literary Digest,* 119 (April 20, 1935), p. 10.

"Dust: More Storms Wreck Destruction in the Southwest." *Newsweek,* (April 20, 1935), p. 11.

"Dust-Storms' Aftermath." *Literary Digest,* 120 (November 2, 1935), p. 15.

"Dust Storm Film." *Literary Digest,* 121 (May 16, 1936), pp. 22–23.

Eddy, Don. "Up from Dust." *American Magazine,* 129 (April 1940), pp. 54–55, 89–92.

"Electrified Dust Storms of the West." *Popular Mechanics,* 49 (February 1928), p. 226.

Ellis, Lippert S. "The Soil Conservation Districts Law in Oklahoma." *Southwestern Social Science Quarterly,* 19 (September 1938), pp. 183–88.

Evans, Morris. "Nonresident Ownership—Evil or Scapegoat?" *Land Policy Review,* 1 (May/June 1938), pp. 15–20.

"Federal Movie Furor." *Business Week* (July 11, 1936), p. 14.

Finnell, H. H. "Dust on the Plains." *Audubon Magazine,* 50 (January/February 1948), pp. 16–19.

————. "Dust Storms of 1948." *Scientific American,* 179 (August 1948), pp. 7–11.

————. "Dust Storms of 1954." *Scientific American,* 191 (July 1955), pp. 25–29.

————. "The Plowup of the Western Grasslands and the Resultant Effect Upon Great Plains Agriculture." *Southwestern Social Science Quarterly,* 32 (September 1951), pp. 94–100.

Fite, Gilbert C. "Farmer Opinion and the Agricultural Adjustment Act, 1933." *Mississippi Valley Historical Review,* 48 (March 1962), pp. 656–73.

Fossey, W. Richard, "Talkin' Dust Bowl Blues: A Study of Oklahoma's Cultural Identity During the Great Depression." *Chronicles of Oklahoma,* 55 (Spring 1977), pp. 12–33.

Fuller, Norman G. "Walter Plagge—One Who Stayed in the Dust Bowl," *Land Policy Review,* 3 (October 1940), pp. 37–39.

Fuller, Varden, and Janow, Seymour J. "Jobs on Farms in California." *Land Policy Review,* 3 (March/April 1940), pp. 34–43.

Glick, George W. "The Drought of 1860." *Transactions of the Kansas State Historical Society,* 9 (1905–1906), pp. 480–85.

Graznak, Michael. "Too Early for Hysteria in Drought-Hit Wheat Areas," *Farmland News* (March 15, 1976), pp. 12–13.

Gray, L. C. "Federal Purchase and Administration of Submarginal Land in the Great Plains." *Journal of Farm Economics,* 21 (1939), pp. 123–31.

"Grasslands." *Fortune,* 12 (November 1935), pp. 59–67, 185–90, 198, 200, 203.

Grimes, W. E. "The Effect of Improved Machinery and Production Methods on the Organization of Farms in the Hard Winter Wheat Belt." *Journal of Farm Economics,* 10 (April 1928), pp. 225–31.

Gropper, William. "The Dust Bowl." *Nation,* 145 (August 21, 1937), p. 194.

Gutherie, John D. "Trees, People and Foresters." *Journal of Forestry,* 40 (June 1942), pp. 477–80.

Hall, William L. "The Shelterbelt Project." *Journal of Forestry,* 32 (December 1934), pp. 973–74.

Haslam, Gerald. "What about the Okies?" *American History Illustrated,* 12 (April 1977), pp. 28–39.

Hatfield, Russell. "People on the Plains." *Kansas Water News,* 13 (1979), pp. 1–16.

Hersig, Carl P. "New Farms on Newly Irrigated Land." *Land Policy Review,* 2 (November/December 1939), pp. 10–16.

Henderson, Caroline A. "Letters from the Dust Bowl." *Atlantic,* 157 (May 1936), pp. 540–51.

————. "Spring in the Dust Bowl." *Atlantic,* 159 (June 1937), pp. 715–17.

Henson, Edwin R. "Borrowed Time in the Dust Bowl." *Land Policy Review,* 3 (October 1940), pp. 3–7.

Hibbs, Ben. "Dust Bowl." *Country Gentleman,* 210 (March 1936), pp. 5–6, 83–87.

————. "Reaping the Wind." *Country Gentleman,* 104 (May 1934), pp. 15, 45, 48.

Hill, Gilbert. "The Plains Will Blow Again." *Science Digest,* 17 (June 1945), pp. 89–92.

Huntington, Ellsworth. "Marginal Land and the Shelter Belt." *Journal of Forestry,* 32 (November 1934), pp. 804–12.

Hurt, R. Douglas. "Dust!" *American Heritage,* 28 (August 1977), pp. 34–35.

————. "The Dust Bowl." *American West,* 14 (July/August 1977), pp. 22–27, 56–57.

Idso, Sherwood B. "Dust Storms." *Scientific American,* 235 (October 1975), pp. 108–11, 113–14.

Isley, C. C. "Will the Dust Bowl Return?" *The Northwest Miller,* 224 (November 20, 1945), pp. 18, 35, 38–39.

Jones, Ewing. "Dust Storms Through the Years." *The Land Today and Tomorrow,* 2 (April 1935), pp. 1–4.

Judd, B. Ira. "The Dust Came at Noon." *The Cattleman,* 60 (May 1974), pp. 64, 66, 74.

Keith, B. Ashton. "A Suggested Classification of Great Plains Dust Storms." *Transactions of the Kansas Academy of Science,* 47 (1944–1945), pp. 95–109.

Kellogg, Royal S. "The Shelterbelt Scheme." *Journal of Forestry,* 32 (December 1934), pp. 974–77.

Lambert, Roger (ed.). "A Texas Rancher in the 1950s Drought." *Heritage of Kansas,* 10 (Summer 1977), pp. 21–31.

———. "Drought, Texas Cattlemen and Eisenhower." *Journal of the West,* 16 (January 1977), pp. 66–70.

———. "Slaughter of the Innocents: The Public Protests AAA Killing of Little Pigs." *Midwest Quarterly,* 14 (April 1973), pp. 247–54.

———. "Texas Cattle and the AAA, 1933–1935." *Arizona and the West,* 14 (Summer 1972), pp. 137–54.

———. "The Drought Cattle Purchase 1934–1935: Problems and Complaints." *Agricultural History,* 45 (April 1971), pp. 85–93.

———. "Want and Plenty: The Federal Purchase Relief Corporation." *Agricultural History,* 46 (July 1972), pp. 390–400.

Lane, Neil. "The Dust Farmer Goes West." *Land Policy Review,* 1 (May/June 1938), pp. 21–25.

Logsdon, Guy. "The Dust Bowl and the Migrant." *American Scene,* The Thomas Gilcrease Institute of American History and Art, 1971.

Lorang, Glenn. "Wanted: Water for the Plains and West." *Farm Journal,* 101 (March 1977), pp. 32B, 33A.

McDean, Harry C. "The Okie Migration as a Socio–Economic Necessity in Oklahoma," *Red River Valley Historical Review,* 3 (Winter 1978), pp. 77–91.

McEntire, Davis. "Migrants and Resettlement in the Pacific Coast States." *Land Policy Review,* 1 (July/August 1938), pp. 1–7.

———, and Whetten, N. L. "Recent Migration to the Pacific Coast." *Land Policy Review,* 2 (July/August 1939), pp. 7–23.

McGinnis, B. W. "Utilization of Crop Residues to Reduce Wind Erosion." *The Land Today and Tomorrow,* 2 (April 1935), pp. 2–14.

McGinty, Brian. "The Dawn Came But No Day." *American History Illustrated,* 7 (November 1976), pp. 8–18.

MacMillan, Robert F. "Farm Families in the Dust Bowl." *Land Policy Review,* 1 (September/October 1938), pp. 14–17.

Malin, James C. "Dust Storms Are Normal." *University of Kansas Alumni Magazine,* 52 (March 1954), pp. 4–5, 54.

_____. "Dust Storms: Part One, 1850–1860." *Kansas Historical Quarterly,* 14 (May 1946), pp. 129–44.

_____. "Dust Storms: Part Two, 1861–1880." *Kansas Historical Quarterly,* 14 (August 1946), pp. 265–96.

_____. "Dust Storms: Part Three, 1881–1900." *Kansas Historical Quarterly,* 14 (November 1946), pp. 391–413.

Malone, A. W. "Desert Ahead!" *New Outlook,* 164 (August 1934), pp. 14–17.

Mollin, F. E. "If and When It Rains: The Stockman's View of the Range Question." Denver: American National Livestock Association, 1938.

Munns, E. N., and Stoeckeler, J. H. "How Are the Great Plains Shelterbelts?" *Journal of Forestry,* 44 (April 1946), pp. 237–57.

Nall, Garry L. "Dust Bowl Days: Panhandle Farming in the 1930s." *Panhandle–Plains Historical Review,* 47 (1975), pp. 42–63.

_____. "Specialization and Expansion: Panhandle Farming in the 1920's." *Panhandle–Plains Historical Review,* 46 (1974), pp. 46–67.

Norris, Ada Buell. "Black Blizzard." *Kansas Magazine,* (1941), pp. 103–4.

"Parched Plains: Experts Survey Drought Ravages as Government Speeds Relief Moves." *Literary Digest,* 122 (August 29, 1936), pp. 7–8.

Pearson, Charles G. "Drama in the Dust Bowl." *Kansas Magazine,* (1952), pp. 94–97.

Pew, Thomas W., Jr. "Route 66: Ghost Road of the Okies." *American Heritage,* 28 (August 1977), pp. 24–33.

Plough, J. S. "Out of the Dust." *Christian Century,* 52 (May 22, 1935), pp. 691–92.

"Pros and Cons of the Shelterbelt." *American Forests,* 40 (November 1934), pp. 528–29, 545–46.

Rasmussen, Wayne D. "The Impact of Technological Change on American Agriculture." *Journal of Economic History,* 20 (December 1962), pp. 578–91.

"Record Survival Made by Shelterbelt Trees." *Journal of Forestry,* 40 (June 1942), p. 456.

Reitz, T. Russell. "A Traveler Sees the Shelterbelts." *Progress in Kansas,* 7 (December 1940), pp. 11–12.

"Remember the 30's, 50's." *The Marketeer,* 9 (April 1977), pp. 10–12.

Richardson, Rupert N. "Some Historical Factors Contributing to the Problems of the Great Plains." *Southwestern Social Science Quarterly,* 18 (June 1937), pp. 1–14.

"Roosevelt and the Drought." *Nation,* 143 (July 18, 1936), pp. 61–62.

Rowell, Edward. "Drought Refugee and Labor Migration to California in 1936." *Monthly Labor Review,* 43 (December 1936), pp. 1355–63.

Russell, J. S. "We Know How to Prevent Dust Bowls." *Journal of Soil and Water Conservation,* 10 (July 1955), pp. 171–75.

Schuyler, Michael W. "Drought and Politics, 1936: Kansas as a Test Case." *Great Plains Journal,* 15 (Fall 1975), pp. 3–27.

Sears, Alfred B. "The Desert Threat in the Southwestern Great Plains: The Historical Implications of Soil Erosion." *Agricultural History,* 15 (January 1941), pp. 1–11.

Seibert, Victor C. "A New Menace to the Middle West: The Dust Storms." *The Aerend,* 8 (Fall 1937), pp. 209–26.

Shaw, Elmer W. "A New Look at the Shelterbelts." *American Forests,* 63 (December 1957), pp. 18–19, 47–49.

"Shelterbelt Plantings for 1938." *Journal of Forestry,* 36 (June 1938), p. 581.

Sherman, E. A. "Saving Our Soil." *Nation,* 36 (April 12, 1933), pp. 401–3.

Sorensen, Willis Conner. "The Kansas National Forest, 1905–1915." *Kansas Historical Quarterly,* 35 (Winter 1969), pp. 386–95.

Stoeckeler, J. H. "Shelterbelt Planting Reduces Wind Erosion Damages in Western Oklahoma." *Journal of the American Society of Agronomy,* 30 (November 1938), pp. 923–31.

Strode, Josephine. "Kansas Grit." *Survey,* 72 (August 1936), pp. 230–31.

Taylor, Paul S., and Rowell, Edward J. "Refugee Labor Migration to California, 1937." *Monthly Labor Review,* 47 (August 1938), pp. 240–50.

————, and Vasey, Tom. "Drought Refugee and Labor Migration to California, June–December, 1935." *Monthly Labor Review,* 42 (February 1936), pp. 312–18.

"The Recent Destructive Dust Cloud." *Science,* 79 (May 25, 1934), p. 473.

"The Shape of Things." *Nation,* 143 (July 1936), p. 29.

"There's Still a Drought." *Business Week* (November 17, 1934), p. 12.

"Topsoil Wind Erosion Is Worst in 20 Years." *The Marketeer*, 9 (May 1977), p. 8.

Traxell, Willard W., and O'Day, W. Paul. "Migration to the Pacific Northwest, 1930–1938." *Land Policy Review*, 3 (January/February 1940), pp. 32–43.

Tripp, Thomas A. "Dust Bowl Tragedy." *Christian Century*, 57 (January 24, 1940), pp. 108–10.

Udden, J. A. "Dust and Sand Storms in the West." *Popular Science Monthly*, 49 (September 1896), pp. 655–64.

VanDoren, Mark. "Further Documents." *Nation*, 142 (June 10, 1936), pp. 753–54.

Van Royen, William. "Prehistoric Droughts in the Central Great Plains." *Geographical Review*, 27 (October 1937), pp. 637–50.

"Wallace Praises Prairie Tree Planting." *American Forests*, 45 (July 1939), p. 377.

Ward, Harold. "Conquering the Dust Bowl." *Travel*, 74 (February 1940), pp. 24–25, 48.

Weakly, Harry E. "Recurrence of Drought in the Great Plains During the Last 700 Years." *Agricultural Engineering*, 46 (February 1965), p. 85.

Wessel, Thomas R. "Prologue to the Shelterbelt, 1870 to 1934." *Journal of the West*, 6 (January 1967), pp. 119–34.

————. "Roosevelt and the Great Plains Shelterbelt." *Great Plains Journal*, 8 (Spring 1969), pp. 57–74.

Whitfield, Charles J. "Sand Dunes in the Great Plains." *Soil Conservation*, 2 (March 1937), 208–9.

————. "Sand Dunes of Recent Origin in the Southern Great Plains." *Journal of Economic Research*, 56 (June 15, 1938), pp. 907–17.

————. "Wind Erosion Endangering Colorado Vegetation." *The Land Today and Tomorrow*, 1 (December 1934), pp. 27–28.

"Will It Be a Dust Bowl All Over Again?" *U. S. News & World Report* (April 26, 1976), p. 69.

"Wind Erosion Damage Reported," *The Marketeer*, 9 (April 1977), p. 5.

Winters, S. R. "New Plants to Check Drought and Dust." *Literary Digest*, 119 (April 13, 1935), p. 19.

Wood, Ivan D. "Irrigation in the Middle West." *Agricultural Engineering*, 38 (June 1957), 418–21.

Woodman, H. V. "Pasture Development in Texas." *The Land Today and Tomorrow,* 2 (March 1935), pp. 7–11.

Worster, Donald. "Grass to Dust: Ecology and the Great Plains in the 1930's." *Environmental Review,* No. 3 (1977), pp. 2–11.

Zingg, A. W. "Speculations on Climate as a Factor in the Wind Erosion Problem of the Great Plains." *Transactions of the Kansas Academy of Science,* 56 (September 1953), pp. 371–77.

Zon, Raphael. "Shelterbelts—Futile Dream or Workable Future?" *Science,* 81 (April 26, 1935), p,. 391–93.

UNPUBLISHED MATERIAL

"A Memorandum on Factors Involved in the Drought Relief Situation." FERA, Divisions of Research, Statistics and Finance, Washington, D.C., January 15, 1935. National Agricultural Library.

"A Report of the Soil Conservation Service to the Secretary of Agriculture on Problems of the Southern Great Plains." C.W.A. Subject File, Documentary Resources. Colorado State Historical Society.

"A Report of the Soil Conservation Service to the Secretary of Agriculture on Problems of the Southern Great Plains and a Conservation Program for the Region." April, 1954. National Agricultural Library.

Adams, Ruby Winona. "Social Behavior in a Drought Stricken Texas Panhandle Community." Master's thesis, University of Texas, 1939.

Alden, Verna Carney. "A History of the Shelterbelt Project in Kansas." Master's thesis, Kansas State College of Agriculture and Applied Science, 1949.

Alsup, Frances McNeal. "A History of the Panhandle of Texas." Master's thesis, University of Southern California, 1943.

Ballantyne, John N. "The Prairie States Shelterbelt Project." Master's thesis, Yale University, 1949.

"Chase County Historical Society Sketches." Kansas State Historical Society.

"Dust Storms of 1911–13 Are Worst in History of Area." *Thomas County Yesterday . . . Yesterday and Today.* Kansas State Historical Society.

Floyd, Fred. "A History of the Dust Bowl." Ph.D. dissertation, University of Oklahoma, 1950.

Green, Donald P. "Prairie Agricultural Technology, 1860–1900." Ph.D. dissertation, Indiana University, 1957.

Grey, Pauline W. "The Black Sunday of April 14, 1935." *Pioneer Stories of Meade County, Kansas.* Kansas State Historical Society.

"History of Smith County, Kansas." Kansas State Historical Society.

Kansas Water Resources Fact Finding and Research Committee. "Water in Kansas: A Report to the 1955 Kansas State Legislature." July 26, 1954. Kansas State Historical Society.

Kincer, J. B. "Man's Responsibility for Droughts in the Great Plains." Paper presented before a meeting of the American Meteorological Society, Pittsburgh, Pennsylvania, October 29, 1934.

"Land Use Survey of the Southern Great Plains; Revised April 1, 1938." Division of Land Economics, Land Utilization Program, USDA, Bureau of Agricultural Economics, Southern Great Plains Region, Amarillo, Texas. Amarillo Public Library.

Lang, James B. "The Shelterbelt Project in the Southern Great Plains—1934–1970—A Geographical Appraisal." Master's thesis, University of Oklahoma, 1970.

Pfister, Richard. "A History and Evaluation of the Shelterbelt Project." Master's thesis, University of Kansas, 1950.

"Report of the Eighteenth Conference of the Regional Advisory Committee on Land Use Practices in the Southern Great Plains Area." Agricultural College, Stillwater, Oklahoma, October 21–22, 1938. Amarillo Public Library.

Sjo, John D. "Technology: Its Effects on the Wheat Industry." Ph.D. dissertation, Michigan State University, 1960.

Stewart, H. L. "Changes on Wheat Farms in Southwestern Kansas, 1931–37." USDA, Farm Management Reports No. 7, Washington, D.C., 1940. National Agricultural Library.

"Summary of Drought Relief Programs, 1932–1952." Washington, D.C., 1954. National Agricultural Library.

"The Dust Bowl: Agricultural Problems and Solutions." U.S. Department of Agriculture, Office of the Land Use Coordinator, Editorial Reference Series No. 7, 1940. National Agricultural Library.

Williams, E. Morgan. "The One-Way Disc Plow: Its Historical Development and Economic Role." Master's thesis, University of Kansas, 1962.

Young, Forrest Albert. "The Repercussions on the Economic System of the Great Plains Region of Kansas of Mechanization of Agriculture." Ph.D. dissertation, State University of Iowa, 1938.

Index

209